The Invasion of Eden

Did our ancestors warn us about ET invasions?

And is history repeating itself in the 21st century?

What people are saying about Paul's Books

Paul Anthony Wallis is one of my best colleagues...very intelligent man and a brilliant author of books. His arguments and proofs are enlightening. Paul has a way of speaking that everyone can understand. I have full respect for him. **Erich von Daniken – Chariots *of the Gods (on the Chariots of the Gods Podcast)***

"Incredible! Paul has written an amazing series and with what timing on this particular topic! Everything in our past has led up to what we are now experiencing and we need to be aware and make mindful choices. It is the most important thing right now." **Laura Eisenhower**

Paul Wallis takes us on a journey that we will never forget. This generation's "Chariots of the Gods!" **George Noory – *Coast to Coast, Beyond Belief***

Fantastic! Enthralling! Paul is a really incredible researcher - one of the most interesting researchers when it comes to paleocontact. **Benjamin Grundy – Mysterious Universe**

"A true gentleman and a meticulous researcher, Paul Wallis brings humility and insight to one of the most contaminated fields of enquiry – 21st century UFO-logy. We highly recommend Paul's extraordinary works of investigation to all those seeking a more enlightened path. History will recognize Paul's important works." **Jaimie and Aspasia Leonarder - The Movie Show SBS**

Paul Wallis is the new Erich von Daniken. A remarkable orator. For anyone interested in the cross-over of religion and ET's, Paul's work is a must-read. **Regina Meredith on *Open Minds GAIA TV***

It is a great pleasure to collaborate with Paul. I love our work together. Paul's contribution is excellent and for me it is important to compare my studies with Paul's! Though far apart geographically we are spiritually close! We are a good team.
Mauro Biglino

I have read many books on contact. Paul's books are of a completely different ilk. They come with so much credibility, so much research, and so much to back them up. To anyone wanting to know about ET's and asking, 'What's the truth?' I would say, 'Read Paul's books!' **Sandra Sedgbeer - OM Times**

"This man is a genius who speaks real truth, with a wealth of real knowledge. Paul Wallis has got his hands dirty and has done his research. He is a real reverend who has put in the work. This is why this guy is my hero!"
Billy Carson on *4Bidden Knowledge TV*

Paul is doing a courageous service [giving] us a new perspective on the creation and engineering of man.
Sean Stone – Actor, Media Host

Paul is an extraordinary researcher who radiates tranquillity, scholarship and courage. Paul Wallis does the footwork, and it really shows in his scholarship. **Revd Dr Sean O'Laoire PhD**

Paul's wit and humility are second only to his deep knowledge of human history and mankind's origins. I strongly recommend his books. **Jay Campbell – Researcher, Bestselling Author**

Paul Wallis expresses the awareness of many cultures in such a personable way. Really fascinating. I have learned so much.
Barbara Lamb – Licensed Psychotherapist

Also by this Author

Escaping from Eden

ISBN: 978 1 78904 387 7

The Scars of Eden

ISBN: 978 1 78904 852 0

Echoes of Eden

ISBN: 978 0 6454183 0 9

The Eden Conspiracy

ISBN: 978 0 6454183 2 3

Paul Wallis Books

First published by Paul Wallis Books, 2024

www.paulanthonywallis.com

Text copyright: Paul Wallis 2024

ISBN: 978-0-6454183-5-4

Design: Tempting_Dezine

The Invasion of Eden

Did our ancestors warn us about ET invasions?
And is history repeating itself in the 21st century?

Paul Wallis

Paul Wallis Books

Acknowledgments

Without the generous encouragement of my beautiful family, Ruth, Evie, Ben, Caleb, Hugo, and Skye, and of my *5th Kind TV* collaborator Anthony Barrett, this book would not be in your hands today. I have to pay tribute to my amazing wife who has provided tireless and patient assistance with this project every step of the way. Thank you My Wonderful!

I am deeply indebted to my fellow explorers, Matthew LaCroix, Richard Dolan, Laura Magdalene Eisenhower, Narine Grigoryan, Hasmik Sargsyan, Billy and Elisabeth Carson, Tony and Leah Aquino, Tara Alamodin and David Lovegrove for their contributions and their encouragement on this research path.

I am continually grateful to all those around the world who read, like and share my material, and all those who have helped to propel www.paulanthonywallis.com the *Paul Wallis Channel* and *5thKind.TV* to even wider audiences.

I would also like to acknowledge my compatriot Ross Coulthart, who, with journalistic integrity and tenacity, has pursued the story of David Grusch's revelations concerning *The Program,* bringing it into the mainstream so that everyone can now consider the implications of what is happening in the corridors of power. In a world where the narrative is so fiercely policed by government and the owners of our mass media, we need more independent journalists of courage like Ross.

I want to pay tribute to Kam and Kalea and all my brothers and sisters in Hawaii, for their resilience through the suffering and loss caused by the terrible Maui fires of 2023. I am grateful for your love and hospitality and my heart is with you as you grieve and rebuild.

As with all my books, you will find that I have conflated various visits, characters and occasions with just enough creativity to protect the privacy of those who have entrusted me with their conversations, to whom many thanks once again.

I dedicate this volume in the *Eden Series* in loving memory of my mother, Brenda Marcia Wallis (1933-2023) without whose sharp mind, inquisitive spirit, and skill with words, I would not be doing the work I do today. Thank you Mum!

INTRODUCTION & CHAPTER ONE

Bitten

International Airspace 2023

I have delayed as long as possible. The creeping sensation has reached the point where the pain of holding it in is greater than the disruption of making a move.

In the world of American military intelligence I am in good company. This has been a year when a struggle over what can be held back and what can be released has been a matter of public scrutiny, most notably in the U.S. Congress.

Leading figures in the Pentagon are clearly concerned. They fear that the repercussions of releasing more information to the public would be simply too unpredictable. Meanwhile, other leading figures in the world of military intelligence have let more cats out of the bag with regard to ET contact in the last few months than we have seen in more than seventy years, in all the time since the whole question of UFOs in Earth's airspace first exploded into the public consciousness.

Because of what has been revealed this year, a number of members of Congress are now on the warpath. They believe that an existential threat to the planet ought to be everybody's business. Their view is that if the Pentagon really does have access to non-human technology and non-human pilots (and yes that means ETs) then this information is far too important to be withheld from Congress.

Now various members are challenging the legality of the Pentagon's secrecy surrounding these matters. They want accountability regarding what the Pentagon has done with the trillions of dollars of public money poured into what it calls *The*

Program. What is the justification for these black budgets? Who or what are we in contact with? What exactly have we reverse engineered? Congress wants to know. In the words of Congressman Tim Burchett, *"We're done with the coverups and we're going to get to the bottom of it, goddamnit!"*

The thing that has sparked this drama between Congress and the Pentagon was a landmark complaint made by an officer of high standing in the military intelligence community, David Grusch. If you haven't followed this story in the news, and you don't know who David Grusch is and have never heard of *The Program,* you are certainly not alone. Beyond a couple of media outlets in the U.S.A. the furore surrounding Grusch's formal complaint has gone largely unremarked upon. This is a mistake. His allegations and the credentials which back him up make David Grusch an historic figure. But I am getting ahead of myself, and I will introduce you to David properly a few pages from now.

For the moment, suffice it to say that David Grusch's testimony, given under oath before the Congressional House Oversight Subcommittee on July 26th 2023, and televised for the world to see, was something without precedent in more than seventy years. For the first time in three quarters of a century, the question of extraterrestrial visitors has become a matter of Congressional concern. In all my life I have never seen official acknowledgement of the UFO phenomenon at this level. It's palpable. Something has to give.

But that isn't my problem. That is not the creeping sensation that is bothering me in international airspace, somewhere over the Aegean Sea. My problem is that I need to pee. However, as much as I would like to relieve myself, I cannot move. It's not the fact that my seat, 66F, the middle seat, in the middle row of the economy section of my transatlantic flight, has sandwiched me between two large, unconscious passengers, who will have to be

14

disturbed or clambered over for me to get to the bathroom. No, the issue is that I am physically unable to stand up, and I don't know why. Looking down at my feet I try to process what I am seeing. I see two feet in matching white Calvin Klein sport-socks. But the feet inside those socks are suddenly far from identical. This is not how I remember them. My left foot is its usual flat fish shape. My right foot, however, is like a water balloon.

If this is the first time you and I have met, then you will be as baffled by this as I was. On the other hand, if you have read my previous *Eden* books, you may have a sneaking suspicion as to what injury this may be. However, this is not that. Suddenly feeling somewhat nauseous, I jab at the call button for the flight attendant. Within a minute the flight attendant has arrived, assessed the situation, and produced a couple of anti-inflammatories. Meanwhile my neighbour in 66E, who happens to be a physiotherapist, gently reassures me that latent injuries have a certain way of flaring up in the pressurised environment of an aircraft cabin. However this is not an old injury lying latent. It's a spider bite. I just don't know that yet. The puncture wounds are still too subtle to be seen and the eerie lake of blackness that will soon replace the sole of my foot won't fully reveal itself for another two days. So we are all in the dark.

"Sir, when we disembark would you like us to assist you with a wheelchair?"

A wheelchair? It will be a poignant enough reunion with my parents without the unexpected shock of seeing their son being ferried through customs in a wheelchair.

"No. That's very kind but no thank you. I think I'll be alright. Perhaps if you can provide me with a robust baggage trolley, I think I should be able to support myself on that."

Once on the ground and de-planed, my trolley strategy proves successful with the result that my parents and I are some distance between the airport terminal and the car park before my mum picks up on the shocking state of my dodgy foot.

As it happens, I am a seasoned traveler, and this is only one of many inter-continental adventures under my belt. In my early years, my travel was on account of matters of international security in an age of terrorism. My father's work in this arena took our family to every continent around the world, a happy by-product of his professional endeavours to make the world a safer place in which to fly. Today my travel is less about geopolitics and more about *exopolitics*. This word refers to the investigation of our place, secure or otherwise, in a populated cosmos, and that is the reason for my flight today. My route into this controversial field was one I could never have predicted, even a few short years ago.

As I write, it is thirty-nine years since I first entered the world of church-ministry, and this year it will be thirty-three years since I was ordained as a priest in the Church of England. My ecclesiastical career culminated in serving the Anglican (what in America is called the Episcopalian) Church in Australia as an Archdeacon – that's one rung down from a Bishop. The critical roles that led me into the world of extraterrestrial contact were those of Church Doctor and Theological Educator.

The work of a church doctor is usually focused on pastoral transitions and appointments, financial and governance problems. However it also, on occasion, requires a surgical response to paranormal activity in churches and parishes. It is work which cannot help but stretch one's worldview a little further than average. In my work as a theological educator I designed and provided in-service training to pastors in the disciplines of the *History of Christian Thought* and *Hermeneutics* – the principles of interpreting ancient texts. These were the responsibilities that led me to engage with a non-human aspect of the human experience in the world of our ancient texts, including in the pages of the Bible. The more attention I paid to the root meaning of key words in those texts, and the more I looked at anomalies in the Biblical narratives, the more I was able to recognize a body of cultural memory of extraterrestrial contact in the deep past. (If that intrigues you, you can follow the details of that journey in my book *Escaping from Eden.*) So although it is certainly unusual for a senior churchman to be a spokesperson in the world of close encounters and Ufology, my work has made this territory unavoidable in my field of study.

If you raised an eyebrow at the subtitle of this book because you find the idea of ET invasion faintly comical, then once again, I have to acknowledge that you are probably in good company. From the *Invasion of the Body Snatchers* to *Mars Attacks* and *Nope*, from *War of the Worlds* to *Independence Day,* popular culture has generally encouraged us to consider the prospect of ET contact as nothing more than fantasy and entertainment. However the ambit of ancestral narratives makes these topics quite unavoidable. In the realm of world mythology stories of invasion are front and centre and, as I have learned through a long hermeneutical journey, extraterrestrial invasion is a thread which weaves through the world's most popular ancient text, the Bible. If you're not familiar with the Bible, or if you have never read it with ET contact in mind, stick with me and be ready to be surprised.

In the next chapter we will travel to a truly historic part of the world, rich with archaeological evidence of humanity's deep past. There we will get up close and personal with a mysterious basalt box, hard copy evidence of our ancestors' recollections of cosmic contact. Together we will examine an anomalous artefact which, though it surfaced more than half a century ago, is only now beginning to offer up its secrets.

However, before I tell you more about this object, and let you know where our journey will take us, I should explain why any of this has any relevance to you and me today. Why should you and I care what ancient elders sowed into the world's canon of origins stories? Why should we care about anomalies in the writings of ancient scribes or the mysterious designs of ancient artists etched into the stone structures of long forgotten cultures? How is any of that relevant today? Don't we have more pressing concerns in the twenty-first century? For a proper answer to those questions, we must begin our journey in Washington, on Capitol Hill, where for the first time since the 1940s Congress is petitioning the Pentagon to disclose what it knows, and has known for eighty years, about ET contact.

Capitol Hill, Washington – Wednesday July 26th, 2023

"My name is David Charles Grusch. I was an intelligence officer for fourteen years, both in the U.S. Air Force (USAF) at the rank of Major and most recently, from 2021 to 2023, at the National Geospatial Intelligence Agency at the GS-15 civilian level, which is the military equivalent of a full-bird Colonel."

"I was my agency's co-lead in Unidentified Anomalous Phenomena (UAPs) and trans-medium object analysis, as well as reporting to UAP Task Force (UAPTF) and...once it was established, the All-Domain Anomaly Resolution Office (AARO)."

"I became a whistleblower, through a PPD-19 Urgent Concern filing in May 2022 with ICIG (the Intelligence Community Inspector General), following concerning reports from multiple esteemed and credentialed, current and former military and Intelligence Community individuals, that the U.S. Government is

operating with secrecy - above Congressional oversight - with regards to UAPs."

(UAPs is today's nomenclature for UFOs.)

"My testimony is based on information I have been given by individuals with a longstanding track record of legitimacy and service to this country – many of whom also shared compelling evidence in the form of photography, official documentation, and classified oral testimony to myself and many of my various colleagues."

"I have taken every step I can to corroborate this evidence over a period of four years while I was with the UAP task force, and to do my due diligence on the individuals sharing it. And it is because of these steps that I believe strongly in the importance of bringing this information before you. I am asking Congress to thoroughly investigate these claims."

How did I get here? A few short years ago my work was a million miles away from Ufology and the halls of political power. My work as a senior churchman immersed me in the sedate world of Christian ministry, serving various communities of faith, and quietly publishing books on Christian spirituality. Today I am known around the world as *"The Paleocontact Guy."* Paleocontact is the theory that in the deep past our ancestors experienced contact with visitors from other civilizations and sowed the memory of that contact deeply into their narrative traditions. It is this research path which has brought me to today's Congressional Hearing. Given everything I have put my name to over the last few years on *5thkind.tv* and in my books in the *Eden Series,* what is happening in Washington right now is something I cannot ignore.

It would be an understatement to say that my life changed dramatically when the first in this series, *Escaping from Eden,* became an international bestseller. On his flagship radio show, *Coast to Coast,* George Noory hailed it as *"This generation's 'Chariots of the Gods.'"* The sequels *The Scars of Eden, Echoes of Eden* and *The Eden Conspiracy,* along with my work on the *Paul Wallis* channel and *5thkind.tv* have led to media commentators such as Regina Meredith of GAIA TV affirming me as *"The new Erich von Daniken."* And just to show that he doesn't take any offence at this, Erich Von Daniken has personally endorsed my work, saying, *"Paul Wallis is one of my best colleagues. I have full respect for him."*

You are now probably thinking that I must be riding high on ego and self-satisfaction to have told you that, but the truth is that I have to really exert myself to overcome my English upbringing and suppress my British reserve to self-recommend in that way. And I can't not tell you, because when Erich Von Daniken spoke those words on his own YouTube channel, for me it was like receiving the affirmation of a parent. This is because I was only eleven years old when Erich first sowed in my young mind all the deep questions concerning human origins which were to lead me first into the world of Christian ministry and ultimately into the world of paleocontact. That is why I am here today, paying such close attention to the sworn testimony of David Grusch, Commander David Fravor and Ryan Graves, Executive Director of Americans for Safe Aerospace.

David Fravor and Ryan Graves are both highly credentialled and experienced naval pilots. Back in 2004, it was David Fravor who commanded the squadron dispatched from the USS Nimitz to engage with the now famous Tic-Tac UAPs, the video footage of which has now been viewed by billions of people all around the world. Who David Grusch is is key to this whole story and is the very reason for this Congressional Hearing. David Grusch has worked at the highest levels of U.S. Intelligence. In his capacity as Lead Investigator for the UAP Task Force in the Pentagon, it was his job to collate information about *"The Program."* The Program is the intelligence community's name for the matrix of agencies, intelligence officers and sub-contractors, devoted to reverse-engineering metamaterials obtained from UFO crash-retrievals. For more than seventy years this work was undertaken with maximum levels of top secrecy. Then in 2019 the Pentagon finally acknowledged publicly the existence of this top-secret program, having been outed by a semi-official leak in 2017, courtesy of former Assistant Secretary of Defense, Chris Mellon, published in the New York Times. Since that time compelling detail has been provided by the former chief of the Advanced Aerial Threat Identification Program, Luis Elizondo, and further corroborated by Alain Juillet the former Chief of French National Security. Yet more detail has been made public by eminent scientists Dr. Jacques Vallee, Dr. Eric W. Davis, Dr. James Lacatski, and Nobel Prize nominee Professor Garry Nolan, all of whom have now spoken publicly about their responsibilities in reporting technical findings evinced from the analysis of exotic materials retrieved from *"off-world vehicles not made on this Earth."*

These extraordinary disclosures led to Senate Briefings in 2021 and Congressional hearings in 2022. Resultant of that, as a senior intelligence officer within the Pentagon, David Grusch was entrusted by Congress to investigate *"The Program."* Is it for real? Are there UFO craft in the possession of top-secret levels of US Government? Of course as Lead Investigator for the *National Geospatial Intelligence Agency UAP Task Force* David Grusch would need access to *"The Program"* in order to fulfil his remit. But he was refused. For all his authority, and even with top-secret SCI clearance – a clearance he still holds – he simply was not allowed to know.

What makes this moment even more significant is the extraordinary way this impasse has been allowed to escalate. Following his being blocked from *The Program*, David Grusch went to DOPSR (the *Defense Office of Pre-Publications Security Review*) and obtained permission to speak publicly about the Pentagon's program of crash retrievals, which he has now done. Though clearly ring-fenced with NDA's and bound by official secrets laws, without naming places and names publicly, although he has done privately, David Grusch has now spoken to journalists in America, concerning the Pentagon's access to *"non-human"* technology and pilots. Just to make it clear, he has done this with the blessing of two key intelligence authorities ICIG and DOPSR.

We now know that David Grusch has spoken to people who have direct knowledge of The Program, as his job required. And forty of those people, under oath, have now corroborated David Grusch's report. Their sworn testimony has provided ICIG with:

- The present locations of retrieved craft
- The names of people who have run *The Program*
- The names of the Pentagon officials responsible for blocking its disclosure
- The details of what was recovered, when and from where.

Just look at those dot points and let them sink in for a moment. This was the information which impacted ICIG.

At the instigation of Thomas Monheim, the Inspector General of the Intelligence Community (ICIG) a letter was then sent out to the various Senate and House Intelligence Committees, communicating that ICIG had determined that David Grusch's claim concerning a denial of access was *"urgent"* and *"credible."* This is why you and I are here today listening to members of Congress as they interrogate three highly credible witnesses. I won't go over the whole hearing but suffice it to say that Ryan

Graves and David Fravor provide first-hand testimony of their encounters as naval pilots with UAPs and make a strong case for their non-human origins, based on their first-hand engagement. These cases are already in the public domain. Still more intriguing is the moment when David Grusch is asked to go further in his explanation of contact, not with craft only but with *"non-human"* pilots. At this point I have to pinch myself that we are in Congress, talking about contact with *"non-human"* pilots of *"non-human technology."* This is one of several moments in which David Grusch replies *"I can't discuss that publicly. I can only answer that in a closed session,"* or *"I can give you an answer but it will need to be in a S.C.I.F."* (That's a Sensitive Compartmented Information Facility.) So clearly, there is something to see here.

"Mr. Graves...I would like to ask; how do you know that these were not our aircraft?"

The question is from Tim Burchett, congressman for the second district of Tennessee.

"Some of the behaviours that we saw: We would see these objects at 0.0 mach...These objects were staying completely stationary in category 4 hurricane winds. These objects would then accelerate to supersonic speeds...and they would do so in very erratic and quick behaviours that we don't...have an explanation for."

Congressman Burchett now turns his attention to Commander Fravor. *"Mr. Fravor, what astonished you the most about the flight capabilities of these Tic Tacs?"*

Again there is no hesitation *"Their performance. Absolute performance!"*

"And [Mr. Fravor] you're not aware of any other objects that anybody in the world has - in this world - that has those capabilities?"

David Fravor replies, *"No. Actually I think it's far beyond the materials science that we possess."*

Why is Congress interested in discussing UAP's or UFOs? In case there should be any doubt, the point of concern is made crystal clear when congressman Andy Ogles asks all three witnesses this pointed sequence of questions:

"Is it possible that these UAPs may be probing our capabilities?

The response is unanimous.

"Yes." "Yes," "Yes."

"Is it possible that these UAPs might be testing for vulnerabilities in our current systems?"

"Yes." "Yes." "It's possible."

"Do you feel based off of your experience and the information that you have been privy to that these UAP's provide an existential threat to the national security of the United States?"

Again our witnesses are in agreeance: *"Potentially." "Potentially." "Definitely, potentially."* As we will see in a minute, *"potentially"* is a significant word-choice.

"In the event that your encounters had become hostile, would you have had the capability to defend yourself, your crew, or your aircraft?"

"Absolutely not." "No."

There is absolutely no mistaking that concern. The presence of *"non-human"* visitors with advanced technology is being framed as an issue of military threat, despite the fact that in all the official reports of UAP encounters filed by U.S. defense, as surveyed in the 2021 Senate Briefing Paper, there is not even a single incidence of *"non-human"* aggression against our planet's military installations or civilian populations. That being the case, why is a non-human presence in our skies and under our oceans being framed as *"an existential threat"* to the United States of America? Never mind the rest of the world!

Of course, purely on the basis of logic, any technology more advanced than our own, any craft that can toy with our pilots in the way reported by David Fravor and Ryan Graves has to be a *"potential threat."* It's only logical. Hence the respective answers of our three witnesses: *"Potentially." "Potentially." "Definitely potentially."* However, given the total lack of aggression over a documented period of eight decades, are our members of Congress right to frame our visitors as *"hostiles"*?

As a student of world mythology, I instinctively turn to the treasure trove of ancestral narrative and ask, *"Did our ancestors have anything to say to these concerns? Did they speak of E.T. invasions? Is such an invasion even credible? Has it already happened? And if so, how would we know?"*

To answer these questions I am going to take you on a global journey. In the chapters ahead we will travel to one of the planet's oldest cultures, that of ancient Armenia, which includes the lands of Anatolia in Modern Turkey. From there we will continue to the Red Sea and on to West Africa, from the Trossachs of Scotland to the foothills of the Rocky Mountains of Colorado and on again to the Hawaiian island of Molokai. From the indigenous narrative traditions of those places we will distill evidence that all is not as it appears in the intersection of government and Ufology.

If we have the courage to listen, our ancestors' stories will open up a whole new world of possibilities for every individual on the planet and for our whole civilization. If you are ready to cross a few cultural boundaries and have your worldview rotate a few degrees, then I invite you to join me on this global tour. In the next chapter, our travels will take us to an ancient basalt box which will confront us with data and designs which pre-date every religion, every wisdom tradition and every national identity that defines the world today. It will offer us a window onto a time in humanity's deep past when we were visited by certain non-human others. I have a feeling this box may be an artefact left behind from the time long ago when we were last invaded.

CHAPTER TWO

What's in a name?

The Bosphorus Strait

The line in the ocean where the Marmara Sea meets the waters of the Black Sea is, for me, one of the great wonders of the world. As guests of Andreas Potamianos, the owner of the Epirotiki fleet, we are sailing on the World Renaissance, a large enough and very capable ship. Yet a boyhood spent bingeing on thrillers and disaster movies means that as I stand on the deck, surveying the uncannily visible line where the two seas meet, I am picturing our beautiful ship suddenly being sucked down into the deadly vortex produced by the difference in density of the two seas.

Of course, there is no deadly vortex and it is perfectly smooth sailing as we continue down the Bosphorus Strait to Istanbul, the great city which straddles Europe and Asia. It is a dramatic entrance to a unique and fascinating city, rich with history. The reason I have brought you here is that while our friends in the corridors of twenty-first century power are framing our visitors as *"hostiles"* menacing our planetary security, just down the road from here, the testimony of one of the world's most ancient cultures paints a far more detailed and layered picture of what contact may have looked like to our distant ancestors. I want to spend a while here letting their perspective pull us into a time long before the language of Ufology, and long before all the taboos around cosmic neighbours which for centuries have been a part of our mainstream culture and many of our inherited religions.

What I want to show you today is on the Asian side of Istanbul, a place six hundred miles to the East of the great city, where excavations for a school have ground to an unexpected halt. Local authorities have summoned archaeologists to the school at

Adilcevaz, which sits adjacent to the banks of Lake Van in the region of Kef Kalesi, where the area has now been cordoned off. Site preparations have revealed mysterious megalithic remains which are going to require careful analysis. This find is part of an archaeological puzzle which has been unfolding in the region since the 1960s. The artefacts being excavated are in a region known for the Urartu culture, which flourished around the turn of the first millennium BCE, as a later offshoot of ancient Sumer. Names, glyphs and symbology adorning Urartu's archaeological artefacts all point in that direction. There is, however, an anomaly here, inviting us to look more deeply. Literally.

When in 2017 the find at Adilcevaz leads a team of divers to begin examining the bed of Lake Van, the sophistication of the megalithic remains which they find submerged in the lakebed, hundreds of feet below water level, has prompted a question raised by other ancient megaliths found in Lebanon, Peru, and pre-dynastic Egypt: Why do the earlier, deeper strata of civil engineering exhibit a level of technological ability which is superior to the more recent layers? What story do these stories want to tell us?

At Adilcevaz the questions run still deeper. Indeed it is the very depth of the lakebed where these megaliths have been found which makes its artefacts so exciting and so important. The depth and the age of the lake provide the anomaly. This is because the highly sophisticated community responsible for the lower strata artefacts was above water-level no more recently than 6,500 years ago. Yet Urartu as we know it dates only to 3,000 years ago. That's quite a disparity. There is a gap in the story.

In 1984 archaeologists Egon Degens, How Kin Wong, Stephan Kempe, and Fikret Kurtman undertook a geological study of the historic depths of Lake Van. It identified a window, a period of three thousand years, during which megalithic structures could have been constructed at a depth just short of a thousand feet below the current water level. That window was between 6,500 years ago and 9,500 years ago, close to the tail end of the most recent ice age, the Younger Dryas Cold Period. This pivotal moment pre-dates all the timelines we have hitherto associated with the Sumerian civilization and its offshoots by thousands of years and pushes us closer to a date associated with the relaunch of Earth's human population, a moment commemorated in a whole world of ancestral narratives.

A question mark is raised over the relationship of the finds above water with the finds below. Is the history of the Urartu culture deeper and more important than we have previously understood? Or were the engineers of Urartu's prehistory preceded, or even tutored, by an even more ancient, pre-Younger Dryas Ararat Civilization? Just how ancient is the history here? How deep does this rabbit-hole go?

There are many reasons why people with an interest in the human story would already have their eye on this region, the lands which comprised ancient Armenia. For instance, in 1998 a team from the University of Norway in As and the Max Planck Institute of Cologne Germany identified and excavated the world's earliest farm at a site in what today is Anatolia, in south-east Turkey. The site at Karaca Dag bore the archeological signatures of a seismic leap forward in humanity's grasp of agronomy. What the archaeological team uncovered was the world's earliest evidence of the genetic manipulation of plants in order to create cultivatable crops, alongside the sudden appearance of animal husbandry. The team, led by Professor Manfred Heun of the University of As, dated this vital node in human history to ten thousand years before present. This is an auspicious date. It represents a crisis in the story of humanity, a pivotal moment in which humanity desperately needed an intervention.

Ten thousand years ago, a drastically reduced human population was emerging from Earth's most recent ice age, a crisis which had brought homo sapiens to the brink of extinction. However, with the advent of crop-farming and animal husbandry, the survivors of the Younger Dryas Cold Period could begin to do something more than survive. Crop farming mean surpluses. Surpluses mean higher density living and all the opportunities of a specialized society. Specialization allows for the exploration of science, art, and all kinds of technological innovation. For all those reasons, people have looked to the Fertile Crescent and to that period of 10,000 years before present to identify the birth of human civilization as we know it, tracing it from its origins at the top of the fertile crescent to the emergence millennia later of the sophisticated metropoles of ancient Mesopotamia - in what today would be the countries of Iraq, Iran and Syria. However, this is where the plot thickens.

If you have already read my book *Escaping from Eden*, you will remember that we need to drive no more than two hours West of Karaca Dag, still within modern-day Turkey, to arrive at a place where this neat picture of human progress becomes disordered. If you think you know where we are headed next, don't skip ahead, because whether you and I have made this connection before or whether this is the first time we have travelled together, we will soon be somewhere quite unfamiliar, reviewing an incredible artefact, one which throws a quite different light on all the angst this year in Washington surrounding national security and existential threats.

Two hours from Karaca Dag is Gobekli Tepe. Here, on a limestone plateau in the foothills of the Taurus mountains, megalithic remains were first unearthed in 1963. Over the last thirty years the incredible structures excavated on this site have raised significant question marks over the claim of Karaca Dag to be the world's first farm. The complex at Gobekli Tepe is extensive. It was not the weekend project of a local tribe of neolithic hunter-gatherers. At the time of writing, Gobekli Tepe is already recognized as extending to an area more than fifty times the size of Stonehenge. By carbon-dating organic matter associated with the megaliths, researchers have been able to ascertain that the site was active before ten thousand years ago, at which point, for some unknown reason, its curators preserved it by very carefully burying it.

Gobekli Tepe is fascinating for many reasons, not least that its function was neither residential nor religious. The central complex appears to be something more like a museum. Moreover it is marked with technical features, glyphs and symbology, which would place it within a global culture. We are looking at the product of a settled and advanced culture. The civilization of which it was just a part pre-dates the pioneering population at Karaca Dag, with Gobekli Tepe being buried around the same time that Karaca Dag was emerging.

This timeline suggests that the civilization to which Gobekli Tepe belonged already existed before the end of the recent ice age 12,800 to 11,600 years ago, during that challenging period, and possibly before. Karaca Dag on the other hand appears to be the re-boot of civilization post ice age. The proximity of Gobekli Tepe to Karaca Dag along with their dovetailing timelines hint at the possibility that the civilization at Karaca Dag may have been tutored by the survivors of an advanced, pre-ice-age civilization, whose culture is memorialized at Gobekli Tepe. All this is challenging enough for our mainstream storytelling of human development. However the finds six hours to the east, at Adilcevaz, take us far beyond the paradigm of human-to-human tuition and into the realm of paleocontact.

One of the most interesting artefacts from the region around Lake Van can now be found in the halls of the Museum of Anatolia, in Ankara. It is a stone box or, more precisely, a giant block of basalt, 4.6 ft by 4.6 ft by 3.6 ft. As The flawless reliefs adorning its impossibly hard, black, volcanic surface have an unsettling effect. The imagery is profoundly foreign, yet strangely familiar to me. The lines are so clean it is as if this giant cuboid has travelled through time to bring us the knowledge of another world, and in a way that is exactly what it has done. But how? Basalt is not an easy stone to work with. In fact it is so hard that iron or steel tools struggle to work with it. In the twenty-first century we would use a diamond-tipped saw to cut it and tungsten-carbide to work with it. How in the world would stone and bronze age tools have produced this?

Just at this moment, though, I am feeling a little distracted. For some reason my right foot is bothering me. It's annoying because I don't want anything to detract from my surveying what has to be the most fascinating artefact yet to have been retrieved from the ancient sites of Kef Kalesi. Pushing this distraction out of my mind, I immerse myself in the basalt box's intriguing designs and drink in their incredible story about Earth's deep past.

What grips me from the start is the familiar symbology of these reliefs. I have seen Sumerian, Babylonian and Assyrian carvings before. They include portrayals of advanced beings, ancient non-human visitors who introduced civilization, as we know it, to our distant ancestors. These figures include Enlil, Enki, Gilgamesh, Oannes and the Apkallu. (If those names are unfamiliar, don't worry. We will look more closely at these figures in a later chapter.) Adorning this giant basalt cuboid in Ankara is a similar figure known as Haldi.

You may have seen the famous Assyrian carving of the Sumerian King, Gilgamesh, five metres tall and holding a grown lion under his arm as if it were a lap cat. Housed in the Louvre in Paris, its original home in the 8th century BCE was the Palace of Sargon II in Khorsabad. Though not as large as Gilgamesh, like him, Haldi is portrayed alongside a lion, standing on top of it in fact, to give us a sense of his power and right to rule. As a native Brit, I recognize the lion. It adorns the coat of arms symbolizing the right to rule over the country of my birth. The meaning of the basalt lion under Haldi's feet is the same. This is Haldi's country.

In the case of the lion-bearing Gilgamesh, his right to rule is explained in the Babylonian and Sumerian stories etched into the cuneiform tablets of those ancient cultures. These written texts describe Gilgamesh as a hybrid person, human but enhanced with the DNA of mysterious beings who came from space and ruled over our ancestors, the Sky People. It would seem that he and Haldi may be of the same stock.

The lion under Haldi's feet connects him with another advanced being from the world of ancestral story, Asherah. She is the female figure, commemorated by the tribes of Israel for her arrival from the stars and her nurture of their ancestors, tutoring them in the secrets of agronomy and civilization-building. She too was associated with a lion, two lions in her case, illustrating her power and prowess. The other symbology on this box tells us even more about what these mysterious beings are offering to their human neighbours, suggesting that Haldi be part of a benevolent visitation like that of Asherah, bearing profound gifts for humanity.

Looking to the right of the box, Haldi stands bearing a cloak and sporting a pair of body-length feathered wings. In his left hand he holds a bowl. The bowl or handbag motif repeats in carvings and reliefs in India, Cambodia, Central and South America. It appears to represent knowledge, science, medicine and technology. His extended right hand presents the offer of a pinecone, a symbol associated with wisdom, medicine, or technology. The pinecone hints at higher powers of cognition, activation of the brain's pineal gland, the ability to attune to higher orders of understanding and higher ways of being. This echoes the Sumerian story of another visitor from the stars, Enki. In the Sumerian cuneiforms, Enki is the one credited with genetically engineering the first human beings, nurturing our beginnings as a species, and fine-tuning our cognitive powers. Is Haldi part of that cadre of visitors?

While Haldi is at ground level, eagles can be seen hovering in the sky which would seem to represent the power and authority to govern. Reflecting the eagles' wings, Haldi's own wings raise further questions. Is Haldi being portrayed as non-human? Or are the wings symbolic of a mastery of the air which our ancestral viewers found difficult to comprehend? Either way, Haldi is represented as being present on Earth as the representative of an authority in the sky. The eagle powers are shown to be in their abode in the sky while Haldi goes about his business on the planet's surface. This scenario echoes in the Babylonian story of an advanced being called Oannes. Like Haldi, Oannes resides on Earth, while advancing human development as an emissary of authorities in the heavens, or what we would call outer space.

Between Haldi on the ground and his feathered counterparts in the heavens, are three devices which give the appearance of ziggurats, being four sided, three-stepped pyramids. Inverted pyramids can also be seen at sky-level. This symbology is instantly recognizable to anyone even glancingly familiar with the carvings and reliefs of ancient Mesopotamia.

"Kef Kalesi is a site that almost nobody is talking about." I am talking with my friend, the researcher and author, Matthew LaCroix, who I first met when he guested for the first time on *5thkind.tv*. *"I really want to make known why it is so important. This huge basalt box emerged from the banks of Lake Van*

decades ago. Yet I haven't found any papers or promotion of this site. When I found it through examining terrain on Google Earth and poring over photographic images from various university departments, I had never heard of it."

I have to agree. In fact it was Matt who introduced it to me while on his research path for his latest book, *"The Epic of Humanity,"* co-authored with our mutual friend Billy Carson. (I wish I had thought of that title!) It is a pleasure for me to sit with Matt and compare notes regarding the cosmological implications of the Kef Kalesi cuboid. I love Matthew's company and his passion for the subject is absolutely infectious, and as he expounds upon various finds in Anatolia, I can feel the momentum for a documentary movie beginning to build. Matt explains how it is that the basalt cuboid has remained so quiet since its re-emergence onto the world stage more than half a century ago.

"I have come to the conclusion that this may be one of the most significant ancient reliefs anywhere in the world. When they found it, they thought this was just another artefact of the Urartu culture. Because of the dating we usually associate with the Urartu culture, the find just hasn't received the attention it truly deserves."

The reliefs on the box certainly tell a rich story. Along with spacefaring technology, advanced knowledge and cognitive upgrades, are other images which suggest meanings both biological and esoteric. Banks of seeds are coming from the sky, and, at ground level are three trees. The balance of the three trees strikes me, as well as the fact that they are each in a pot being carefully cultivated.

"Compare it to the Tree of Life mural at the Ashurbanipal Library in Iraq, where the Epic of Gilgamesh was uncovered. The similarities are astounding."

The correlations are clear. Dated in the conventional way as a later Sumerian offshoot, the parallels would be, although interesting, not as significant. If, on the other hand, this basalt box belongs to the period of what is now on the bed of Lake Van, then the box's information becomes hugely significant. It would represent a missing link in the story of human civilization, a vital link in the chain, far earlier than the dates historians generally associate with ancient Sumer.

As Matt introduces me to the region's archaeological artefacts another layer of questions is firing in my mind. Before paleocontact and before the theology that got me there, my first love was languages and linguistics. Through school and university studies in Britain, Italy and Brazil, I was lucky enough to learn French, German, Portuguese, Italian, Latin and ancient Greek. I have always been fascinated by the history of language and I know that Armenian is one of the world's most ancient languages. Unlike modern hybrid languages like English or Italian, which are great, great, great grandchildren of the proto-Indo-European language, Armenian stems from the very root of that language.

Evolutionary Biologist, psychologist and co-director of the Max Planck Institute, Russell Gray, and Professor Quentin Atkinson of the University of Auckland are both experts in the theory of language evolution. Together, they have produced evidence through their joint research which dates the Armenian language to the ninth millennium BCE. This means that as a language Armenian carries truly ancient information within its vocabulary, and from a part of the world with some of the earliest evidence of human habitation and civilization. So we are in territory where we should fully expect to find some of the very earliest artefacts of human culture and mythology. For all those reasons the name *Haldi* grips me. Stay with me for the next few pages and you will see why the name of Haldi gives us every reason to suppose that the carvings on his box are indeed ancient memories of an extraterrestrial conquest.

In this part of the world Haldi is the great patriarch and *"god."* His looming presence throws a different light on another patriarch, a figure who is foundational to three major world religions, Islam, Christianity and Judaism. I am talking about Abraham. The Genesis narrative introduces the reader to Abraham in a place called *Ur* (an Assyrian word part which means *high place*) and from among a people called the *Chaldes*. Let me tell you where this word comes from. *Chaldes* is an English transliteration of the ancient Greek-language rendering of Genesis in the Septuagint. (The Septuagint is the authoritative ancient Greek translation of the Hebrew Scriptures.) The Greek spelling is *Xaldaion*. It refers to a people group – the people of *Xaldi*. The relevance of this is in how this word from antiquity pronounced.

The *X* of *Xaldaion* is pronounced as a voiceless velar fricative *CH* as in the Scottish *loch*. In proto north-west Semitic, the ancestor of Biblical Hebrew, this sound was indicated by the letter *H*. Writers in the classical period preferred to indicate the same sound with a *CH*. So for simplicity, because they all make the same sound, for the next few paragraphs, I will represent the *X/CH/H* sound as an *H* to highlight the phonic connections between the words we are about to survey.

When Abraham emerges from the high place or *Ur* of the *Haldaion*, he carries what would appear to be a summary form of the ancient Sumerian stories. For this reason and others, scholars have long believed that Abraham's famous journey probably began in Mesopotamia in what is modern Iraq. This has been the received wisdom since Henry Rawlinson of the East India Company first proposed the idea in 1862. However no location for the high place of *Haldaion* has ever been proven. In the 1920s British Assyriologist Sir Leonard Woolley proposed that *Ur* of

Haldaion was a place known today as *Al-Muqayyar* in Iraq. However none of the names associated with that place bear any linguistic relation either to the Hebrew, *Ur* of *Kasdim,* or the Greek *Ur* of *Haldaion* (in the Greek.) Yet, the absence of better identifications, Leonard Woolley's suggestion as to which *Ur* we may be talking about has been left largely unchallenged in all the time since.

There is also a problem with the longstanding identification of Abraham's *Ur* of the *Haldaion* with the Mesopotamian country of *"Haldea,"* and that is that Mesopotamian Haldea did not actually exist at the time of Abraham. Haldea in Mesopotamia existed only from the 10th to the 6th centuries BCE. The latest possible date for Abraham is at least one thousand years too early, some time in the 18th century BCE. So, given that mismatch of timelines, we have to ask, where were the lands of Abraham's people of *Haldi* if not in lower Mesopotamia?

To locate the people of *Haldi* in the right historical timeframe for Abraham we have to look further back in time and a little further north-east to the lands of ancient Armenia. Here we find people, whom classical authors called the *Haldoi* and whom they honoured as the first culture to work with iron tools. Classical authors referred to the region as *Haldia* – a word which connotes the land of *Haldi*. Do you see where this is going?

If Abraham's *Haldaion* are really the *Haldoi* of *Haldia* and are the people of *Haldi*, then these equations would locate Abraham in the regions which today constitute Anatolia in the lands of ancient Armenia. I might be sticking my neck out here but can there really be any doubt that we are talking about the *Ur,* the high country, of our friend *Haldi.*

If what I am proposing is right then we need to revise how we understand the father figure of the world's three major monotheistic religions: Islam, Christianity and Judaism. And as we think through the implications, it will inform a new understanding of the Abrahamic stories of non-human visitors and their intervention in the story of human origins.

Before we get there, however, just to add to the weight of what I am suggesting, take note that the book of Genesis reports that Abraham relocated from the high place of the *Haldaion* to a city called *Harran*. If Abraham was moving from *Ur of Haldi* in ancient Armenia towards his ultimate destination of Canaan, then en route he would have arrived at a city called Harran, which is only two hundred and forty miles to the south-west of the region of Lake Van. The city of Harran is between four and four and a half millennia old. So it exists within the required timeframe to be Abraham's Harran. This two-hundred-and-forty-mile distance is a substantially more likely journey than the eight-hundred-mile trek required by Henry Rawlinson's identification of Ur of the Chaldes with Al-Muqayyar in modern Iraq. In fact, for a personal point reference, it is about the same distance my maternal grandfather walked one hundred years ago to get from Brynmawr in Wales to Buckinghamshire, England, in search of work. The proximity of Harran, Anatolia, to Lake Van is telling.

To add yet further weight, generations later, we hear Abraham's grandson, Jacob, referred to as an *"Arammi obed"*, a phrase which means a *lost, displaced, wandering, migrant* from the land of *Aram* (and it is the Bible's only use of the designation *Arammi*.) To my mind, these linguistic correlations give us numerous, and compelling reasons to look to the high country of Ararat, in ancient Armenia, and wonder if that might be the place from which Jacob's historic *Arammi* lineage emerged.

This has been a dense foray into names and places and I should thank you for sharing this deep dive with me. In the next chapter I am going to tell you why any of this matters.

CHAPTER THREE

Gods and Doors

Ankara, East Turkey 2024

As we probe the correlations of *Haldi, Ur, Aram* and *Harran*, the numerous linguistic connections raise an interesting possibility: the earliest Biblical motifs of invading sky armies, the *Seba Hassamayim, El Elyon the senior elohim, El Shaddai the Destroyer, Yahweh, El of Ekron, El of Egypt, Chemosh, Dagon* and all the invading *elohim,* may be rooted in the experience of ancestors beyond lower Mesopotamia, in the high country of Haldi to the north, in the region of Lake Van in modern-day Anatolia. If that is the case then it follows that these literary motifs can be found in their ancient and maybe original form, etched with diamond grade precision into andesite reliefs and carved blocks of basalt. If the connections I have flagged are valid, then Haldi's fascinating box gives us eyes on how mysterious elohim and Sky Armies appeared to Abraham, his ancestors and his descendants. What did these beings from the sky look like? Were they human, human-like or something else. And if they were not terrestrial, who and what were they?

For all these reasons I am listening with interest as Matt LaCroix raises the whole question of chronology and asks just how old that box might be. Armenian cultural memory extends far into history and to what in school we considered *"pre-history."* In modern Armenia, the people refer to themselves not as *Armenians* or *Haldoi,* but as the *Haya* of *Hayastan,* the country of the mythical King Hayk. In the cultural canon of ancient Armenia, the figures of Hayk and Haldi are very strongly associated with each other. This is especially intriguing because Hayk's associations take us in two directions, firstly into the territory of human origins and secondly into the realm of outer space.

Hayk is a giant, like Gilgamesh the giant hybrid king of Uruk in the Sumerian canon, or Minos the giant hybrid king of the Minoans. Just as Minos is connected with the *"gods"* of Olympus, and Gilgamesh with the Anunna or people from the *"sky"*, so Hayk has his own connection with the stars, specifically with the constellation of Orion. The significance of Orion for paleocontact is that it is one of three regions of space consistently identified by indigenous cultures around the world as the place of origin of our ancient visitors from the stars. Those three regions are Sirius, the Pleiades and Orion. So, in one way or another, all three mythical figures, Gilgamesh, Minos and Hayk, represent a bridge between humanity and a more advanced race of beings from the stars. These figures offer us a connection between humanity and the life of other worlds. Also like Haldi, Hayk is associated with weaponry and conquest. He is also honoured as a once or twice great-grandson of *Hapet* (*Japheth* in Hebrew) the son of the Biblical Noah - Ziusudra in ancient Sumerian, Utnapishtim in Babylonian, and Atrahasis in Akkadian. This would make Hayk one of the fourth or fifth generation of humanity's repopulation of the world, post-cataclysm. Since the book of Genesis identifies Ararat as the *high place* on which the Ark of Noah came to rest after the great deluge, and from which point the human race was relaunched, this claim of a local patriarch only three or four generations removed from Noah, is a significant claim indeed. It is a claim of great longstanding because, since the very earliest of times, the Armenian people have held great pride in their country as the place from which human civilization was rebooted.

To find another vector pointing conspicuously towards ancient Armenia we can turn to Norse mythology. Thirteenth century Icelandic historian, Snorri Sturluson, argued from the ancient Viking sagas that Odin, and the people who migrated with him through Russia to conquer Germany, Norway, Denmark and

Sweden, originated in *Tyrkland* in *Asaland*. That's ancient Armenia. This was the high country in which Sturluson located *Asgardr*, the fortified mountain city of the *Aesir*. In his work, *The Prose Edda*, Sturluson writes:

"Odin was accompanied by a great army of old and young, men and women, carrying their treasures with them. Through whatever lands they travelled, the things spoken of them were so great that they were regarded more like gods than men."

We might pause to ask what the *"things so great"* flowing from ancient Armenia must have been to have appeared god-like to people scattered in other places. Sturluson's translators use phrases like *"glorious things"* or *"glorious exploits"* which give the idea of unfathomable courage and amazing achievements and abilities. In his prologue to *The Prose Edda* Sturluson gets into more detail, specifying Odin and his wife's powers of precognition and other *"magical"* abilities. Other phrases in his history of the Norse kings suggest that Odin's kind were either superhuman or not entirely human, referencing the group as, *"Odin...and all the Aesir with him, and many other people."* The *Aesir* of the Norse sagas were unmistakably superior to regular human beings, and yet the narratives show them to be emphatically flesh and bone people of some kind. (For more on this topic I would recommend *Viking Superpowers* by my friend and fellow-researcher David Lovegrove.)

Elsewhere in *The Prose Edda* Sturluson recalls the unusual attractiveness and intelligence of Odin's family and retinue, again presenting the Aesir and Odin's retinue as if they were members of an advanced civilization, emissaries of a society whose base was hidden by fortifications in the mountains of ancient Armenia. Their superiority over regular human populations is on full show when, with total ease and apparently without any resistance, Odin annexes lands in modern Germany, Norway, Finland, Sweden and

Denmark, and then divides the lands among his sons. Sturluson adds: *"[Odin's] families became so extensive...that their Asaland (Asian) language, became the mother tongue over all these lands."*

Here, Sturluson is pointing to evidence for Odin's *"exploits,"* outside the sagas. In fact he is highlighting evidence which can be seen to this day in the close linguistic correlations between Nordic languages and the language of ancient Armenia, which Sturlusson references as *Asaland,* part of ancient Asia. By pointing to objective, real-world evidence Sturluson is making it crystal clear that, by his calculations, the sagas of Odin in particular are to be read not as fiction or fable but as fragments of cultural memory. As to the detail of that memory, some scholars suggest that Odin's story may be that of a common era, fourth to sixth century king, associated with historic migrations of that time, whose narrative was then overlaid with the tropes of ancient mythology. On the other hand, in his final work, the pioneering Norwegian researcher, Thor Heyerdahl, presented another view, namely that Odin may have been a real king from around two thousand years ago at the time of Roman expansion. For a point of comparison, in a later chapter I will show that the Biblical story of Abraham is probably the conflation of a Hebrew origins story with a far more ancient human origins story. I wonder if that same kind of fusion is what we are looking at in the sagas of Odin, an iconic patriarch whose story taps a cultural memory that is in reality thousands of years old.

In the ages since, the case for Sturluson's geo-linguistic connection has been strongly argued by eminent academic figures, Swedish geographer Philip Johan von Strahlenberg, Norwegian Professor of Linguistics, Georg Valentin Morgenstierne, and Swedish historian Sven Lagerbring. In the 21st century, Professor Gregori Vahanyan has demonstrated in detail the deep linguistic affinities of Swedish with the language of ancient Armenia. In short, we have data from a whole cluster of academic fields pointing to ancient Armenia as a significant source of language, cultural thought, technological advancement, and human migration. Certainly, an uncanny concentration of archaeological firsts point to ancient Armenia as a powerhouse, stretching from the common era to many thousands of years before present:

- One of the world's oldest languages
- Some of the earliest artefacts of human settlement
- Some of the earliest evidence of building with stone
- The earliest artefacts of agronomy
- Early iron technology referenced in classical literature
- The earliest physical artefacts of iron technology
- A close claim to the Noah story of catastrophe, external interventions and repopulation
- Dating of archaeological artefacts which could precede the most recent planetary cataclysm, the Younger Dryas

Quite a cluster! The mythological correlations are also numerous:

- Armenia as the Earthly point of origin of the god-like people of Odin, the Aesir
- Haldi as the bringer of agriculture and civil engineering
- Abraham and the *Ur of Haldi* in ancient Armenia
- *Harran* in ancient Armenia
- Abraham's grandson, the migrant from *Aram*
- Noah and the *high place* of Ararat where the Ark ran to ground.
- Home of Hayk the two-or-three-times great-grandson of Noah *via* Noah's son Hapet (*Japheth* in Hebrew)
- Possible home of Abraham the seven-times great grandson of Noah
- Hayk as a bridge to the stars of Orion

These associations are rooted in rich mythological soil and I have to wonder if in the Viking sagas of Odin, and the archaeology of Hayk and Haldi, we really are tapping a vein of cultural memory relating to key moments of non-human intervention in the story of human progress? Could this be why Haldi's basalt box is adorned with images of advanced beings, along with emblems of science, medicine, a generous spray of seeds and the three trees of life? Aren't these precisely the kinds of symbols we would expect to see if the memory being depicted were of an external intervention, a biological reseeding and cultural rebooting of a chaotic or devastated world?

At this point, you might want to ask me to pause and take a breath. You might be itching to say,

"Paul, get real for a moment! External interventions? Tutors from outer space? How is any of this possible in the real world? Don't the laws of physics rule it out?"

"If even our closest interstellar neighbours were to watch a civilization on Earth approaching some catastrophe or facing extinction through something like a major ice age, surely the limits of the speed of light mean that they would be watching these cataclysms unfold thousands of years after the event, by which time it would already be too late to intervene!"

"Don't the vast interstellar distances render the whole idea of cosmic assistance physically impossible?"

In a later chapter, answers to these precise questions will be put forward by authoritative voices in the world of astrophysics and military intelligence. Specifically, we will hear from four significant scientists deeply involved with The Program. Together, they will invite us to think again about the nature of space, and not on the basis of theory only. These privileged friends are in possession of exotic materials, whose objective properties speak directly to the question of meta-space, wormhole, or sub-space travel. Meanwhile, our ancient Armenian ancestors had their own explanations to offer regarding the logistics of interstellar travel. And let's not presume that our ancestors' explanations must be nothing more than metaphor and fable, because the modalities they suggested have evinced massive investments by U.S. military intelligence for more than thirty years. In that same period NASA has committed billions of dollars to testing these ancestral explanations. As to the nature of these technological modalities, they too are etched in the basalt designs on Haldi's mysterious box.

"Just look at those ziggurats, Paul!"

Matt is highlighting an aspect of the scene adorning Haldi's basalt box, and our attention is focused on three, four-sided, three-step pyramids, positioned side by side, each with an open doorway at ground level.

"What we are looking at is a triptych door! You can see later iterations of that in Da Vinci's Last Supper, or the ancient Peruvian carvings at Machu Pichu, and in in other esoteric art besides."

I have to concur. This motif of the triptych doorway is something I recognize immediately from my forty-two years of immersion in Christian theology. In the tenth chapter of the Gospel of John, Jesus describes himself as a *"doorway"* or *"gate."* A little later in chapter fourteen Jesus refers to himself as embodying *"the way, the truth and the life"* through which people can find their connection with the Source, which Jesus refers to as *"Father, being the one in the Heavens."* (Gk. *Pater eimwn o en tai ouranoi.*) In this way the Gospel writers present Jesus himself as triptych doorway, a threefold and therefore perfect route to a higher way of being.

The same symbology can be found incorporated into church architecture. If you stand in the nave of an eastern orthodox church building and face liturgical east, between you and the inner sanctuary is a threefold doorway. The doors comprise part of the iconostasis, a decorated screen embellished with iconography of Jesus and ascended saints. The three doors are positioned symmetrically, the north door to the left, the south door to the right and the Holy Door in the middle through which the heavily robed priests enter and exit in their mysterious liturgical dance. Together, the doorways and the dance dramatize the mystical sense of connection between the visible realm and the spiritual realm, the Earthly and the heavenly. Put simply, the triptych doorway represents our access on Earth to higher wisdom, higher dimensions and to the eternal realm.

Haldi's triptych predates the church and gospel examples by thousands of years, drawing on a source which predates every religion on the planet today. It just happens to be via Christianity's appropriation of this older symbology that I am quickly able to recognize this layer of meaning in the doorways of Haldi's ziggurats and I enthusiastically explain the parallel to Matt.

"Exactly, Paul! They are doorways to your highest power, to your own ascension."

Without taking anything away from this esoteric reading, there is also a more concrete interpretation of those ziggurats. It relates to the Babel story of Genesis 11 in the chapter immediately prior to the appearance of Abraham in the Bible. The story of the tower in Genesis 11 and its counterpart narrative in the Sumerian *Enuma Elish* confirm each other as accounts of travel between the Earth and the heavens. Genesis claims that the purpose of the *Bab-el* (a word whose root meaning is *power gate*) was to *reach the heavens*. The Enuma Elish specifies that its purpose was to dispatch *observers* from the Earth's surface to their *stations in the stars*. Can you say the word *stargate?*

Like the Vimanas, the airborne temples, and the flying cities of the Indian Vedic texts, images of ziggurats and the built ziggurats themselves, may be the evocations of ancient spacefaring technology. Indeed the association of ancient stargate technology with the Shinar Plain, now in modern Iraq, has been a matter of serious interest to U.S. military intelligence community since the beginning of the millennium - more of which in a later chapter.

In my book, *The Eden Conspiracy,* I describe an artefact retrieved from an archeological dig at Tel El Farah in ancient Samaria. I argue that this object, called a *naos*, represents the device which enabled Asherah of Hebrew memory, the ancient teacher, to travel to the lands of the Levant from her celestial home, a planet orbiting a star in the constellation Pleiades. Just as *Babel* is a linguistic hint of portal technology, so the *naos* is an archaeological hint of the same thing. Haldi's presence on Earth appears to relate to his colleagues' presence in space via the intermediate technology of those ziggurats. Might they be a reference in relief to the same kind of technology referenced in a carved naos at Tel El Farah and a Bab-el memorialized in Biblical and Sumerian texts?

Matt continues, *"This pattern of four-sided inverted pyramids is global. I have tracked it to Ethiopia, Saudi Arabia, Bolivia, Cambodia, Turkey, Sumer, and other places besides. The triple doorway is everywhere too. It speaks about balance, harmony and the pathway to human progress."*

Ascension themes can be found in the foreground of the relief as well. Balance and harmony are reflected in the correlation of

Haldi on earth with his airborne counterparts, the ziggurats reaching to the heavens and the inverted ziggurats reaching to the Earth. The same balance reflects in the calm figure of Haldi on the right, presenting offerings to the left, while his mirror image on the left presents his identical offerings to the right. This visual symmetry suggests balance and communication between left brain and right brain, masculine and feminine, yin and yang. It hints at enhanced vision, intuitive knowledge and activation.

To break it down, the scene on the Kef Kalesi box presents Haldi as:

- A bringer of balance, (the symmetry of the images)
- A bearer of life and health (the trees of life)
- A giver of agricultural science and wellbeing (the seeds)
- A carrier of advanced knowledge, medical science and higher technology (the bowl)
- An activator of higher cognitive powers (the pinecone)

Of course, a relief is a two-dimensional projection. So I can see the connections of above and below, and I can see the lateral movement between left and right. What that leaves to be inferred is how you and I, the observers, relate to what is happening in front of us. Are we receiving that pinecone? Is the cognitive upgrade represented by the pinecone, for you and me? Is this image evoking the memory of a past upgrade of humanity or is it the offer of something ongoing? Where exactly are you and I in that picture?

The scene appears calm peaceable and suggests a benevolent intervention rather than an invasion. However, truth is often a many-layered thing. On the one hand texts etched into andesite and basalt at various sites in the region associate Haldi with symbols of life and ascension. On the other hand, the temples in the region dedicated to Haldi, present him surrounded by impressive displays of weaponry. So, just as Odin was simultaneously handsome and wise, magical and god-like and at the same time was a military force impossible to resist, perhaps Haldi, who conspicuously hailed from the same region as Odin, was not one hundred percent peace and ascension.

As a linguist at heart, I am always drawn to the root meanings of a narrative's key words, so I can't ignore that the etymology of the name *Haldi* takes us to an ancient Armenian word which means *"taken"* or *"conquered."* Whatever the subtleties of who or what Haldi really was, the culture most closely associated with him name him as the great conqueror, which, for all his advancement, would seem to imply that Haldi's arrival was an invasion.

If my paleocontact interpretation of the codes on Haldi's basalt box seems like a stretch to you, the reason I feel comfortable laying it on the table this way is that it is only one iteration of a story which repeats in surprising detail across a great many of the world's indigenous narratives. Our ancestors' vision of extraterrestrial contact was diverse. It included elements of invasion and colonization but also included the memory of help, nurture and salvation from disaster through the intervention of visitors from the stars with advanced technology. From the Nordic countries to the Middle East, from aboriginal Australia to southern Africa, our world heritage is rich with stories such as these.

That being so, why is our twenty-first century Western view so lacking in this background? Why, when we come to Washington in the twenty-first century, are we hearing our political leaders frame the possibility of ET contact solely in terms of national security. Why is that our go-to? Why are members of the U.S. Congress invoking the language of war as they consider the implications of an advanced technological presence in our skies, a presence which in the modern era would appear not to have done us any harm?

I put this question to Nick Pope in a conversation we recorded together for *5thkind.tv*. Nick is the perfect person to answer this specific question for me because, for more than twenty years, he served the British Ministry of Defense as a UFO investigator. His job was to catalogue all UFO encounters with British defense units and report on the implications of those encounters for national security. Nick is a supremely grounded individual, an incredibly thoughtful speaker and, by dint of his privileged position with the Ministry of Defense, his response is informed by his personal access to decades of privileged information. He says,

"The reason we use language like 'threat' is to get buy-in, whether it's [from] the Senate Intelligence Committee, the Armed Service Committee, or whatever it is. You will get more funding if you play up the threat aspect."

Funding. Of course.

Of course, in the absence of transparency there can only be conspiracy theory and conjecture. What we are seeing today is the beginning of a move by Congress to rap its hand on the Pentagon's door and demand disclosure concerning the modern UAP phenomenon and its significance to the United States of America. As an Australian, I am interested in its significance to the whole world. Right now we are in a pause, a momentary silence, awaiting the next move. How will this conversation play forward? Will it be a conversation about cosmic family and our identity within that as the people of Earth? Will it be an acknowledgement of ancient tutors, such as are referenced in countless indigenous narratives around the world. Or will the prospect of a populated universe continue to be framed purely as a military threat, as Nick Pope suggests, for the sake of funding? Given the size of the universe there's no end to the budgets that could be justified that way.

One of the concerns among observers is this: In the 2000s we saw in the *"War on Terror"* how public and Congressional anxiety surrounding potential threats was leveraged by the powers of the day. In brief the U.S. government response (and it echoed around the world) was to increase the scope of its own unaccountable power in the name of *"emergency measures."* Black budgets for military mushroomed, in which climate, according to five successive incomplete audits, the Pentagon consistently managed to mis-account for $1 billion per year, every year, and hold 60% of its assets unaccounted for. In the same period, government has been unable to account for $9 trillion of the public's money. Democratic freedoms were wound back through measures such as the *"Patriot Act"* in the U.S. paralleled by new *"sedition"* laws in Australia. If anxiety about domestic or international threats can be leveraged like that, what might the powers do with public or congressional anxiety about an external, planetary threat?

Speaking to the United Nations General Assembly in 1987, President Reagan wondered aloud whether an *"alien threat"* might unite the human family and bring us all together. Sadly, in light of the *"War on Terror,"* a war with no defined enemy, no geographical limits and no timeframe, it is easy to see how such a planetary threat could be leveraged in another way entirely. It could be made the pretext for even more *"emergency powers"* extended to unaccountable layers of government, and even more democratic freedoms subtracted from the people, all on the grounds of national, no, *planetary* security. It isn't hard to imagine. Anxiety does not make for the best decision-making. Collective anxiety, even less so.

This is why I argue that the longer view provided by ancestral story in verbal and visual form is so helpful. Their long-lived narratives allow us to take several steps back, get a distance from the angst of today's political machinations and begin to get a better perspective. They give us a whole story with a beginning, a middle and an end, to fill out our picture of possibilities. It is for these very reasons that Hawaii is now calling me. It is why in the next chapter you and I will be traveling from Ankara to the beautiful Hawaiian island of Molokai. There, on the green mountainside we will sit at the feet of a cultural elder and drink in the wisdom of indigenous memory which is thousands of years old. It is a canon of stories which may do more than alter your and my perspective on Congress, the Pentagon and UAP's. If we listen deeply, we will be inspired to ask some deeper questions of ourselves, questions concerning who we are as a species, who our visitors may have been in the past and who they may be today. We may come away wondering precisely by whom or what we are being governed.

It is a shame to leave Ankara and say goodbye to Kef Kalesi's basalt box because I know I have only scratched its surface, and I am already calculating how soon Matt and I might be able to get back to this place and dig deeper into what this amazing part of the world has to offer. Though I would love to stay longer, I don't want to be away from my family too long, and right now I am feeling strangely restless. For some strange reason the sole of my right foot is beginning to itch.

CHAPTER FOUR

Scales and Feathers

Molokai – Hawaii – January 2024

"This is sap from the green papaya." Kam moves thoughtfully and I watch him closely as he lights up an unfamiliar cocktail of ingredients on a pot in the campfire and then uses the sap to form a paste with the ash.

"If you had been bitten at home, we would expect to find the elements you need for healing in your home environment. But there will always be something healing in the environment around you, so I will give you what we have here."

I wish I could be more helpful to my hopeful healer, because what I now recognize as a spider bite has probably accompanied me to at least three countries over the last few days. Nevertheless, with my hosts Kam and Kalea, I feel like I am in safe hands. Kam is a devout guardian of ancient knowledge regarding the native flora and fauna of Molokai. He is from one of the *"Old Families,"* and ancestral wisdom flows in his blood. From his birth Kam's Kupuna elders have carefully nurtured in him the higher cognitive powers which they believe are the gift of all human beings. Kam's energy and enthusiasm tell me that he finds real joy in his responsibility to nurture these gifts and curate all the wisdom of his Molokai ancestors.

Far from the popular beaches and other tourist spots, I am staying as a guest on a secluded property in the island's interior. Here I am feeling less self-conscious about my injured foot because in this moment shoes are the exception. It is late afternoon, what the French call *"la crepuscule,"* that liminal moment just before the arrival of dusk. When Kam speaks of finding healing in the environment, I can well believe him. This place is lush and green and the more closely I look, the more fruits and vegetables I can see growing in every corner of his property. Nature is certainly generous here. The company around our conversation pit is warm, and I feel privileged to be with my new friends at their ancestral home. The information Kam is going to share with us throughout this book is centuries, and maybe thousands of years old.

Molokai is tropical and lush, rugged and mountainous. On the surface it is an island idyll, but the sadness of its history is never far from the surface. As Kam introduces me to the local history, I hear a litany of invasions and conquests. Kam's first eye-opener for me is that Hawaii is not American by choice. To settle any doubt about this, the journal of Hawaii's last queen re-emerged in 1993 to tell its story to the world, after nearly a century in hiding. As a prisoner in her own palace, Queen Lili'uokani described her deep anguish as she found herself overthrown and forced to sign deeds of surrender, ceding her kingdom to the USA, only then to be arrested and imprisoned, since her very existence was now counted as an act of treason against the Pax Americana. The coup was the work of the leading American businessmen, the children and grandchildren of the Christian missionaries of a previous generation.

Eager to cash in on Hawaii's abundant natural resources, these powerful businessmen were somehow able to obtain the armed support of an American warship to facilitate their seizure of Hawaiian government. Once Queen Lili'uokani had been apprehended, forced to abdicate and then placed under arrest, Sanford B. Dole, the owner of the leading American corporation, promptly installed himself as Hawaii's new Governor General and immediately set about reorganizing the country to the benefit of his fellow businessmen. And, yes, if that name sounds familiar,

Sandford Dole is the historic patriarch of Dole canned pineapple and fresh fruits. This was how Hawaii began its life as an American territory. Before the Americans, it was the British and before the Brits it was the Tahitians. Before the Tahitians it was the Polynesians, and before them the mysterious Menehune. Before the Menehune it was the even more enigmatic others whose true nature we will come to a few pages from now.

"I was not taught any of this history in school," says Kam. *"Our American textbooks taught me that Queen Lili'uokani gave her kingdom to America, voluntarily. The US government believed that the queen's personal account of what really happened had been destroyed never to see the light of day again. When her diary resurfaced in 1993 it was a powerful moment for Hawaii. It prompted an official apology from the US government under President Clinton, and really sparked today's campaign for sovereignty and independence."*

I can see why it would. Kam is proudly Hawaiian and he delights in his ancestral home. As he applies the paste to my spider-bitten foot, he recounts another story, one that has been in his family for generations, maybe for thousands of years. There is something profoundly comforting about having a person tell you a story. It evokes a feeling of childhood and somehow, I think we all are hard-wired at a deep level to find it healing. It is why in my workshops I always tell stories to my clients. Stories about others help us to see ourselves. Stories from other times and other cultures often help us to get a new perspective on our own, which is why you will find a treasury of stories throughout this book. Plus I enjoy storytelling. I think it's in my Welsh genes. Today though the boot is on the other foot (so to speak) and I am enjoying being pampered and informed by my generous host. As he works his medical magic, I invite Kam to tell me more about the intuitive approach he is now bringing to treating my arachnid injury. Is it innate or has his knowledge been passed down through some tradition of initiation? Kam likes the question and, as he leans into his answer, I can see that he is happy to have an eager student.

"When a baby is born its manawa is open. A Kupuna (traditional elder) can help a person keep it open as they grow into adulthood."

This is a story I am familiar with, not from any prior foundation in indigenous Hawaiian wisdom, but from my travels in France, Central and South America, Greece, Western and Southern Africa, and Aboriginal Australia. If you have read my previous book *Echoes of Eden*, you will know that I love to sit at the feet of indigenous elders and shamanic healers, drinking in the wisdom of traditional cultures around the world and I find that their stories and ceremonies overlap in remarkable ways. They echo each other's themes of contact with cosmic visitors in the deep past, some who came to exploit or enslave us, and others who came to nurture and upbuild us. Their narratives recall a time when our ancestors enjoyed higher cognitive powers than we do today and their ceremonies invoke protocols for reactivating higher levels of consciousness and intelligence. In short, from the Cathars in mediaeval France to the Nangas and Sangomas of today's southern Africa, indigenous traditions have long curated a working knowledge of what Kam calls an *open manawa*.

As a parent of young children I have another reason to take what Kam is saying seriously. Parents, the world over, marvel at their babies' phenomenal innate ability to intuit, perceive and absorb information. Any attentive parent will be able to tell you about flashes of far sight, future sight and empathic connection which they have witnessed in their children when they were very young. Unfortunately, many of our cookie-cutter conventions of education and employment, along with the general busyness of life, seem to have a way of dimming all that bright potential, reducing our inborn genius to the level of industry-ready workers. Indeed, according to our previous government in Australia, this is the ultimate goal of a university education, not that our children should be maximally intelligent, curious, conscious and aware, rather that our children be *"industry-ready."* My friends Kam and Kalea have a different vision of humanity. Their ancestral lore is that of the indigenous people who populated Hawaii in the ages before the Tahitian arrivals of a millennium ago.

"Our ancient priestly lines come from Eden, which we call Havilau. Our princes travelled up the Nile and into Africa. At that time our people had higher powers. Maka hi'e hi'e was normal."

This word *Maka hi'e hi'e* means *remote viewing*. This is a claim shared with the Mayan Popol Vuh which also speaks of a time when our ancestors' vision was not *"limited."*

"Our ancestors had more than head knowledge or intellectual understanding. They could access a wider field of information. This knowledge we call Mana'o. And this deep kind of connection with knowledge we call Na'au."

I am taking notes as fast as I can. *Mana'o* for higher knowledge or knowing. *Na'au* for an interior, intuitive or higher way of knowing. I know those concepts. Moreover to my linguist's ear those Hawaiian words sound strangely familiar. As I mentioned before, the progress of language and the roots of words have always fascinated me. If for a moment we allow ourselves to regard the Greek *s* as a style or suffix, like the Italian *-o* or the Latin *-us*, then the phonetic proximity of *-na'o* and *Na'au* to *No-u* is arresting. The pronunciation is extremely close and the phonetic similarity is significant because of the closeness of their meaning. On the island of Molokai *Na'au* means knowledge that engages mind, heart and higher consciousness, a kind of knowing which draws effortlessly upon a field of information sown into the fabric of the cosmos. *No-u* means exactly the same thing in the ancient language of Hermeticism, the esoteric wisdom tradition which emanated from the priesthoods of ancient Egypt and fused with ancient Greek mysticism and spiritual thought. Its powerful fusion influenced Christianity in its early centuries and found a home in the mystical theology of the eastern churches.

Kam's familiarity with the ancient knowledge carried by his pre-Tahitian ancestry, his facility with shamanic protocols, and his belief that higher cognitive faculties are latent within all of us from our beginning as babies, all ring notes with me, not only because of my contact with other cultures, but because of my academic background in hermeneutics – the principles of interpreting ancient texts.

Hermeneutics was a vital part of my job for the thirty-three years I spent in church-based ministry. As a theological educator, I designed and provided in-service training for Christian pastors in the principles of hermeneutics. When analyzing an ancient text I would teach my students always to ask, *"What kind of literature is this? How do we work that out? And how do we respond to the text once we have worked that out?"* This is generally called *form criticism* or as I would call it *form analysis*.

Secondly, we would ask, *"Is this the original version of the text? Are there clues of an earlier form in wider literature or within the text itself? If what we have in front of us is not the original telling, what was? And if what we have differs from the original, why does it differ? What was the meaning of the original source?"* This is called *source criticism*, or as I would teach my students, *source analysis*.

Thirdly, and fundamentally, we would ask, *"What do the words mean?"* To peel back layers of cultural accretion and religious assumption, I would always direct people to the root meanings of the key words. How does the text run if we substitute the familiar translations with root meanings? At times we would be confronted with words in the Hebrew texts of the Bible which had no etymological history at all and consequently no known root meaning. Sometimes this lack of story behind a word is evidence that we are looking at a loan word, a foreign word which has arrived wholesale from an unknown source to find its home in a new language.

As I look at *Na'au* and *No-u* I wonder what might be going on here. The idea that young children have an open *manawa*, and a natural facility in *Na'au* and *Maka hi'e hi'e* are ideas referenced in Mayan story, as I said before. The idea is also present in the Bible. *Maka hi'e hi'e* is demonstrated by Jesus in John's Gospel, by the prophet Micaiah in I Kings 22, and is present in the origins story of the prophet and seer Samuel.

Tutored by an elder to keep his *manawa* open, Samuel learned as a boy to hold his mind in that liminal space between wakefulness and sleep, in what today we would identify as the brain's theta frequency, in order to be maximally receptive to far sight, future sight, higher cognition and remote communication. Samuel's mentor, Eli, recognized that it was when half asleep that the young Samuel had begun to receive remote communication. Accordingly he told the boy to go back to bed, get back into that liminal state and say, *"Speak, Sir. Your servant is listening."* The take home from this story is that if I wish to heighten my Na'au as an adult, uninitiated into the ways of shamanic acuity, the best thing I can do is to recognize how the theta state feels, learn to hold my mind in that liminal space between sleep and wakefulness and then give attention to the things that come to my mind when in that state. For Samuel this approach would ultimately prove vital to his career. His role in adulthood would be to broker remote communication with an ancestral overlord no longer living on the planet's surface. But that's for a later chapter.

As a father I recognize the innate ability of young children in their theta state through my own experience of parenting. I have seen evidence of an *open manawa* and heard the *Na'au* in the mouths of my children. I have witnessed *Maka hi'e hi'e* in my children and have experienced enough of it myself to be deeply interested in how you and I might nurture the higher cognitive powers which so many indigenous traditions agree are the natural inheritance of all human beings.

This appetite to reawaken abilities, which any and every pastor and priest aspires to do in their ministry practice, is just one of the many reasons I am happy to be sharing this moment with my new friends Kam and Kalea. I am on a learning journey and a healing journey too. Ancient Hermetic practitioners would see these two paths as one. They would call it *ascension,* and my eastern orthodox friends would call it *theosis.* To put it in personal terms, I am eager to make the most of this life and to learn as much as I can in order to have and to share the best human experience possible. However, putting my personal agenda aside for one moment, the reason I have brought you with me today to this hillside property on the island of Molokai is that, embedded within the traditional narratives that Kam curates, are ancient accounts of ET contact, cultural memories of our distant ancestors' experience of colonization by cosmic neighbours, and I want to share with you how they relate to what is happening today.

At a time when members of the U.S. Congress are raising red flags, asking if we are facing an *"existential threat,"* I am happy to be in a place where the atmosphere is less charged, where the knowledge is of a full story arc. I don't want a canvas full of anxieties, what-ifs, speculations or nightmare scenarios. What I am looking for is a wider perspective, the before, during and after of our ancestors' recollections of extraterrestrial visitations. If we had contact in the deep past, what was it like? How did it play out? And in the times when we were invaded, what exactly happened?

"We have been invaded many times," Kam tells me. *"In the time before the flood my ancestors came here from Africa, but their story did not begin there. Our genealogy connects us with our wider family in Alaska, the Philippines, Central, South and North America and Australia. We all know that our shared ancestry goes back before the time when Earth's continents took their current shape. Our common ancestry came from somewhere else. We, the plants and animals, all originated far, far away in Makali'i."* (*Makali'i* is the Hawaiian word for the constellation Pleiades.)

This isn't the first time I have heard this theory. It is there in Plato in his *Phaedo* and *Timaeus and Critias*. It is there in the scientific language of Panspermia, a theory embraced by authorities in DNA research, including Francis Crick the Nobel Prize winner and co-discoverer of the double helix of DNA. Francis Crick argues that the timeline for planet Earth was simply not long enough to have allowed for the emergence of the ordered patterns of terrestrial life's genetic coding. On this basis he argued that all life on Earth must have originated off planet in deep space. Perhaps, he argued, the genetic code for biological and sentient life is as much a part of the normative conditions of the cosmos as the properties of light and gravity. Accordingly, whenever that coding lands in an amenable environment – meaning a planet with water – it will generate forms of life similar to the ones with which we are familiar on planet Earth.

The Zulu people of southern Africa carry a beautifully cinematic version of this story in which the embryonic forms of Earth's animals travel through space in gestation pods and land on planet

63

Earth, where the lifeforms continue to develop until ready for birth and release. The first pod to burst open belongs to the progenitor of humanity, Unkulunkulu, who then opens the other pods, releasing fish into the waters, birds into the air, forest creatures into the forest and the animals of the plain into the savannah. It is an unusual story to invent. Why would animals so holistically embedded each in their own ecological niche, and whose terrestrial behaviours would have been studied intimately by their Zulu observers, be associated with outer space? And why would this African repeat around the world and as far afield as Hawaii?

When Kam names *Makali'i* he is using the Hawaiian word for the stars of the Pleiades and I am not surprised when he mentions them. Nor am I surprised that they would be associated with an indigenous story of human origins. On my research path for my book *The Scars of Eden*, I found myself listening to an Australian pastoralist, Blair, whose family is deeply associated with the stories of their North American ancestors. Blair's family heritage is Cherokee, and for generations his family has curated their people's stories of beginnings, in which their ancestors, living in what is now Tennessee and North Carolina, were visited by beautiful beings who came from the Pleiades, sat with their elders and taught them about the flora and fauna of their land, how to live in balance with it, which plants were good for food, which were good to avoid, which were good for medication and which for higher consciousness. These tutors from the stars taught them new technologies, stories about the wider cosmos and gave them the wherewithal to enjoy a fuller experience on planet Earth. And then, just as mysteriously as they had arrived, they departed in craft which looked like eggs, and returned to their Pleiadean home.

Similarly, when the British began their conquest of Australia, the colonizers asked the original Australians how they had come to be on a land mass so far from anywhere else on the planet. Aboriginal Australian's connection with the land is profound. We now know their presence on Australian soil to extend back at least sixty-thousand years, and possibly as far back as one hundred and twenty thousand years. Even in the 1700's the British invaders quickly became aware of how strongly the original Australians identified with their lands and regions. So they were somewhat surprised when the local elders replied by pointing to the stars, and in particular to the stars of the Pleiades.

Across the Tasman Sea the original people of New Zealand, the Maori, celebrate the appearance of the Pleiades in the night sky every winter. Their appearance evokes joy, peace and gratitude that we are watched over by the eyes of the god, *Tawhirimatea'*. The Maori name for the Pleiades is a contraction of *Nga Mata o te Ariki Tawhirimatea'* – which means *"The Eyes of Tawhirimatea,'"* implying that it is from the Pleiades that we are lovingly watched over.

For all these reasons, when Kam says that *Pa'an* (genealogy) connects his Hawaiian ancestors with Native Americans, Aboriginal Australians and islanders around the world, I don't find that hard to believe. The shared stories of those people groups would seem to confirm it. In my book *The Eden Conspiracy* I show that this Pleiadean connection is there in Hebrew heritage too. The associations are always positive, to do with loving oversight, support and nurture of humanity on planet Earth.

But I have a question for Kam.

"The Pleiades is a big region of space. Is there a particular star within that constellation that your ancestors named as the place of human origins?"

"Yes," he says. *"We call it Mai'a."*

I can hardly miss the phonetic similarity of this Hawaiian word to the name *Maya* by which Kam's distant cousins on the Yucatan Peninsula referred to themselves. If they too saw their deep ancestry as emanating from that same star in the Pleiades, of course they would be the *Mai'a* people.

"Our Pleiadean helpers taught us a way of life that was harmonious. We lived in balance with the animals and plants. Because our higher abilities were activated, because we each knew what the other was thinking, and could feel what the other was feeling, there was no place for deception, separation, and competition. The only way was the way of harmony. We had no hierarchy so there were no masters to slave for. Instead we served one another and found a way of living which benefitted all. This is how it was in the beginning, before were invaded by the Mo'o."

The *Mo'o*, the *Kauna*, the *Ahumanu* and the *Anunu* refer to ancient invaders who were not human. They were reptilians. Some were also feathered. This is an odd pairing of descriptors. Reptilians or lizards don't have feathers, do they? At least they didn't when I was a boy. However, in the twenty-first century things have changed. In the 1970's as I traversed the grades of primary and high school the giant lizards of pre-history were brown-skinned, bald, slow and unintelligent. I am speaking, of course, about the world of the dinosaurs. Since my childhood, though, dinosaurs appear to have evolved. In my formative years, a Diplodocus was an animal so poorly designed as to have been unable to support its own weight and which would therefore spend its adult life consigned to swamps, where the water and marsh could help support its enormous, unwieldy frame. By contrast, today's sauropods can not only support themselves, they can rear up on their hind legs in order to make a display or reach the top branches of the tallest trees. The T-Rex has made similar leaps forward. Recent paleontological finds have revealed that their babies, at least, were not brown and bald after all. They were feathered reptilians.

Prior to these more recent discoveries a feathered reptilian or, to use the traditional nomenclature, *Feathered Serpent,* was the language of mythology and fable. It had no place in the real world. In the mythological world, the Chinese have the *Ikyuchu* and the *Kiucedra* – feathered fire-breathing serpents. The Welsh have the *Draig Coch*, the Georgians the *Kholkhis*, the Spanish and Portuguese the *Coca* – all feathered serpents or dragons. Kam's Mesoamerican cousins had *Kukulkan* a.k.a. *Ququmatz* a.k.a. *Quetzalcoatl*, also feathered serpents. The ancient Egyptians remember *Akhekh*, and the ancient Hebrews in their early stories of Yahweh (originally written without vowels as *YHWH*) speak of leathery skin, a long snout, fire-breathing nostrils and flight feathers. Get the picture? We are talking about dragons. Why would so many disparate cultures independently dream up the same random mixed creature? What does that improbable correlation mean?

"The Mo'o did not come from the Pleiades. They came from the North Star," Kam continues. *"When they came, they brought a different attitude. They altered the energy of our world so that it was better for them, but worse for us. They lowered the oxygen in the atmosphere to dim our higher cognitive abilities and dumb us down."*

This motif is an intriguing echo of the Mayan story, in which the Feathered Serpents release a vapour over human populations to suppress the humans' higher cognitive abilities, specifically to dim their remote viewing, turn off their telepathic connection and shut down their future viewing – all the things that Kam has told us lead to a better human experience and a more harmonious way of life. It is curious to me that the stories of Molokai and the Maya both speak about a modification of the atmosphere to diminish

67

human intelligence. It is an odd idea for ancient human societies to dream up independently of each other – unless of course we really are speaking about a shared collective memory.

In my childhood, it was initially nothing more than a conspiracy theory disregarded as the idle chatter of journalists on slow news days. The idea that lead in the atmosphere from the idling engines of millions of gas-guzzling road vehicles was brain-damaging our children was dismissed as irresponsible fearmongering. That is until the mounting body of unbiased, independent, scientific research proving the connection was made so public that its implications could no longer be avoided. The science was there and report after report proved it. Our children really were being made less intelligent and more aggressive through inhaling lead fumes from an atmosphere saturated with the exhaust fumes of our vehicles.

So, today, when I hear Mayan stories of brain damaging vapours, or the ancient story told by the Efik people of Nigeria about powerful non-human overlords releasing devices to toxify the environment to make humans more manageable, the correlations make me sit up and pay attention. This evening in Hawaii, Kam has shown me that the same memory is carried in the *Mo'olele* – the tradition of his elders.

"The Mo'o turned the world upside-down. Whereas before we lived in harmony and served each other, the Mo-o brought hierarchy. The Anunu brought greed."

Here I really catch my breath. I hate to keep interrupting Kam's amazing story, but this word *Anunu* cannot go without comment. Just like Kam's ancestors, the ancient Sumerians also spoke of reptilian beings. In their ancient carvings, some of these entities were portrayed as feathered. The narratives of the Sumerian cuneiform tablets, and those of their daughter cultures, the

Babylonians, Assyrians and Akkadians, speak of the *Anunna* or *Anunnaki* who, like the Hawaiian *Anunu,* conquered and manipulated our human ancestors. The fact that the Sumerian and Hawaiian names are practically identical, *Anunna, Anunu,* cannot be overlooked, and neither do the similarities end there, as we will see in the chapters ahead. Both groups are remembered as giants. Both are remembered as invaders and overlords. And when we get further into the Hawaiian narrative, we will find that both the Hawaiian Anunu and the Sumerian Anunna shift human culture in suspiciously similar directions.

The names of the *Anunu* and *Anunna* find other phonetic echoes around the world. The Akkadian word for these colonizers from the sky was *Anunnaki*. Biblical narratives recall a similar sounding group, the *Anakim*, a race of gigantic people descended from *Anak*. Similarly, in Greek legend *Anax* was the progenitor of the *Anactorians*, also giants.

Another name Kam introduces me to is that of the *Ahumanu*. We will discover how they fit into the picture in a later chapter. For now I will just mention that their name is very fitting for an invading force with no fellow-feeling for human beings. For *Ahumanu* think *inhuman*. The *Mo'olele* describes the Ahumanu as powerful beings with wings. This aspect repeats not only the narratives of the Feathered Serpents of the Bible and the Mesopotamian narratives, but also the art installations of ancient Mesopotamia, the winged gods in the artistic canon of ancient Egypt, and of course the carvings and reliefs of Haldi in Anatolia.

However, more arresting than the phonetic and visual memory of the Mo'o, the Anunu, the Ahumanu and their international equivalents, is the cultural memory of what these ancient invaders did when they got here, and that is where we are headed in the next chapter. With everything happening this year in Washington with members of Congress getting jittery about *"existential threats"* from *"non-human biologics"* I want to know how our ancestors understood similar threats in the deep past. What was the nature of those threats from the sky? What was it that our ancestors saw? What did our ancient visitors intend by coming? And what happened next?

CHAPTER FIVE

Contracts and Cultural Shifts

The Red Sea

I am half-way between the boat and the shore when a sudden moment of panic grips me. As the waves rock me backwards and forwards it suddenly occurs to me that I have somehow failed to factor into this activity my inability to swim. I am not totally incapable in the water but the shore suddenly looks a long way away. What was I thinking?

Until this moment it has been an idyllic day, spent deep-water line fishing, and filming schools of flying fish as they glide above the warm, millpond waters of the Red Sea. I guess I just got caught up in the moment when our group-leader suggested we all jump in the water and carry our catch of Sea Bass to the shore for a lunchtime barbecue. A decade from now my Paralympian friend Greg Hibberd will teach me to swim properly, but in this moment, I am a poor swimmer, feeling slightly panicked and swiftly calculating whether it will be safer for me to return to the boat or continue with the group. I decide to continue. The group around me all appear to be confident swimmers. Surely one of them can come to my aid if the distance proves too much of a stretch for me. This proves to be a good call. Nothing untoward transpires and before too long I am once again on dry land, standing on the hot sandy beach, soaking in the middle Eastern sun, relaxed and ready for lunch. Ironically, on my return to the boat, I discover that the water is so shallow I can walk half the distance to the boat before I even need to swim. So my stress was all for nothing.

The reason I am in the waters between Egypt and Saudi Arabia has to do with international security, although I am just a hanger on. My dad is the one doing the work. Our time today is a break in that work, a brief pause to enjoy a few hours of leisure with Dad's international posse of colleagues, who have gathered here to develop better systems to keep the world's citizens as secure as possible as they fly around the globe. This is a good spot to choose, auspicious even. Because this sandy shoreline and this body of water are places of phenomenal significance to the human story. We wouldn't be the homo sapiens we know ourselves to be without what happened on these very shores thousands of years ago. According to our Babylonian forbears, it was here that humanity's ascent from among the rank and file of the planet's fauna to become our world's alpha species began. And it was here that the mysterious catalysts of that progress were first witnessed. It is a moment carried deep in our cultural and genetic memory, told and retold by cultures all around the world.

For a couple of examples of that retelling within our own popular culture, consider the moment in *The Jungle Book* when Mowgli leaves his innocent life among the animals of the jungle. Enticed by a beautiful girl, he enters the protection of the human settlement, where he will now learn the ways of civilization. Similarly when a wild, and originally hairy Tarzan meets the civilized Jane, romance quickly ensues. Tarzan's beautiful consort leads him away from his life in harmony with the animals of Africa to Jane's home, a place of modern sophistication where he can learn to enjoy the fineries of civilized living. These resonant stories of Rudyard Kipling and Edgar Rice-Burroughs respectively are a retelling of a sub-story within the oldest written narrative known to humanity – the Epic of Gilgamesh. In the epic's clay pages Enkidu is the primitive human, living innocently in harmony with nature, and Shamhat is the beautiful female who builds a relationship with him through offerings of enticing food and drink. This may be the first iteration of the well-worn proverb concerning the way to a man's heart. Shamhat introduces Enkidu to high-density, city living and to all the comforts of civilization. The story of Enkidu is just one of the ancient Sumerians' literary references to the great leap forward which took their ancestors from a low-density society of foragers and hunter gatherers to a sophisticated expression of metropolitan life, equipped with all the accoutrements of agronomy, civil engineering and city living. A more detailed explanation is given in the literature of the Babylonians, a daughter culture of the ancient Sumerians.

In the 3rd century BCE, the Greek-Babylonian priest Berossus encapsulated the Babylonian explanation of their Sumerian ancestors' sudden progress. In so doing he was addressing a question which to this day has fascinated anthropologists in the field of human origins. How was it that the progenitors of the ancient Sumerians, living at the top of the fertile crescent, first made the great advance from subsistence-living to crop-farming, managing surpluses, creating writing, record-keeping, legal systems and the hierarchy to police them, money and banking and the powers to control them. How, seemingly from out of nowhere, did the Sumerians suddenly possess all the devices for managing larger urbanized populations?

According to Berossus' account of the Babylonian explanation, the catalyst to this rapid and holistic shift was an external intervention. He writes that an intensive education in the management of denser populations was provided to the proto-Sumerian matriarchs and patriarchs by a kind of being so different in physical appearance to its human students that Berossus describes it as *"an extraordinary monster,"* noting that it was first witnessed where I am standing right now on the shores of the Red Sea. Named *Oannes* and the *Apkallu*, these mysterious beings were somewhat similar to humans, the same basic morphology, but larger and somehow fish like. They were part of a mysterious matrix of *"authorities in the sky"* and were able both to live underwater and operate on the planet's surface.

Does that sound familiar? Remember how Haldi was portrayed as active on Earth's surface, while the feathered powers occupied their stations in the skies above? This indicated that Haldi was merely the emissary of a whole spacefaring civilization. It is the same scenario here with Oannes and the Apkallu, notwithstanding their use of an underwater base, to which they would return, so Berossus says, every day at dusk.

Berossus writes: *"Its entire body appeared like that of a fish, and underneath the head was a second one like that of man. Its appearance is still remembered and depicted to this day. This being taught people writing, geometry, science and technology of all kinds. He taught us how to cultivate grains and harvest fruits. He taught civil engineering for city-building and showed us how to establish legal systems. In short, he provided everything that constitutes the marks of a civilization."*

Oannes and the Apkallu are also referenced in the Sumerian *Epic of Gilgamesh* and the Sumerian *Adapa, the Purification Priest of Eridu*. Though remembered as advanced teachers, the most vivid element of Berossus' account is that first impression by the Red Sea and their bizarre physical appearance – like a human but not human, like a fish but not a fish. I wonder if we may be eavesdropping on the first witnesses' struggle to discern between an unfamiliar type of clothing and an unfamiliar type of body. Indeed, there are ancient Mesopotamian reliefs which depict the fish-like elements as apparel and the figures themselves as distinctly human-looking. Yet the first impression as recorded in print by Berossus pulls no punches as to how alien these beings seemed when first encountered by their human neighbours.

The story of Oannes and the Apkallu is a layered story. It differs from other ancient intervention stories from around the world in a number of significant ways. For instance, oral traditions from Aboriginal Australia, Native America, Southern Africa and South America suggest an intervention in our even deeper past, an intervention which taught human beings the art of living harmoniously with nature, and which provided them with an understanding of ecosystems, sustainable farming, sustainable patterns of hunting and fishing, as well as a dietetic and medical appreciation of the surrounding flora and fauna.

The syllabus brought by Oannes and the Apkallu with its overlay of systems of money, banking, record-keeping and law-making, and recipes for beer, sounds a somewhat different note. Their intervention carried the proto-Sumerians well beyond the ambit of ecology and sustainable farming, into the realm of specialized society, civil engineering, complex civics, and economy-building. This syllabus is not about nurturing harmony with the environment, it is about redefining the environment, canalizing the landscape, engineering new eco-systems and moderating the larger populations those systems can now support.

In the twenty-first century, we can only read the Oannes story of social development in the light of everything that has happened since. In that light we might pause and take stock of this metamorphosis. It is a picture of a human population extracted from their native countryside and corralled into cities. Here the people find themselves under the power of a body of laws which have empowered a ruling caste to criminalize and police the grassroot caste. Here they must negotiate a world of employment for money and patterns of banking and record-keeping, which together have created a new social matrix of debts and debtors. And so, from out of nowhere, the dark spectre of poverty and dispossession now hangs over the general population in a way they could have hardly imagined prior to their urbanization – and all this in the context of a society which was newly able to produce surpluses on an unprecedented scale. The incongruity of this polar shift to greater corporate wealth alongside greater insecurity and debt at the grassroots is a pattern we will return to in a later chapter. In the light of present realities, informed by where these shifts have ultimately led, we just might want to question whether this Apkallu syllabus was one hundred percent altruistic.

In a similar vein, Kam explains that when the Mo'o changed the environment it was to advantage themselves as the proprietors of human society at the expense of the human population at large. Though the Apkallu transformation of human society brought previously unimaginable progress and prosperity to the ancient Sumerians, there is certainly a hint of Mo'o-style *"convenience"* in the Sumerian-Babylonian story too. There is an unmistakable light and a shade to the Apkallu syllabus and as I sit today, listening to Kam on the island of Molokai, I am deeply impressed by the correlations between the Mesopotamian accounts of colonization by the *Anunna* Sky People, and the ancient Hawaiian stories of conquest by the *Anunu.*

"The Anunu brought greed. In our language when we think of the Anunu we think of the 'Greedy Ones.' They viewed our land's resources with greed, and they taught us to do so as well, to divide and compete with one another. They set us against one another. Instead of living in harmony and looking after each other we were taught to be self-centred and to fight each other as rivals."

"When the Anunu came, they took the land on which we had all lived freely and privatized it. Our traditional patterns of living on the land and parceling it out were regarded as meaningless. Ownership by the Mo'o took priority. They divided the land, which they said was now theirs to distribute it as they saw fit,

marking the corner of each boundary with a four-sided pyramid to show who now properly owned the land."

"The Anunu introduced money and with it they enslaved us. The moment we agreed with the value of their tokens, which they called 'money,' that was the moment we lost our freedom. The whole meaning of our lives changed on that day. Now everything we did had to be for the purpose of acquiring the tokens, which the Anunu provided."

"In a later time a colonizing king called Pa'auhau ruled over us. He built on these principles of hierarchy and law. He is the one who introduced the Kapu system of laws. Break them and you would suddenly feel the authority of the hierarchy. Keep breaking them and you could be taken away by the hierarchy and killed. That was the law."

My initial reaction on hearing this story is to think what a bleak picture Kam has painted. Yet, when I stop and think about it, the dark threat of violence from the hierarchy is the real basis of law and order in every country in which I have lived. After all, if I refuse to pay what the hierarchy demands of me, in taxes, stamp duty, rates and land rents, then the hierarchy will send people stronger than I am to overpower me, forcibly extract me from my home, physically separate me from my loved ones and my possessions and lock me in a cage until my *"debt"* is paid. If I break the laws that the hierarchy has agreed for me, exactly the same threat of violence hangs over me. In a few countries around the world, if my crimes are violent enough, the hierarchy can do all the above and then kill me. Thankfully, I am glad to say that the places I have lived in have been, for the most part, happy, peaceful law-abiding communities. So it is easy for me to forget that even in the most civilized of settings the state's maintenance of good order and our familiar way of life rests ultimately on a threat of violence from the hierarchy towards the general public should we ever trespass.

It is only on the rare occasions, for instance when the general public protests an unwelcome exercise of power by the hierarchy, that the privilege of violence-without-consequence which belongs uniquely to the state, becomes manifest. When this happens in a peaceful, law-abiding country, such as all those in which I have lived, it comes to most people as a deep shock. So much so, that people struggle to believe what they are seeing, even when caught on camera. I speak from personal experience. So, on deeper reflection, perhaps I can't be too horrified when I hear of King Pa'Auhau or his Mo'o predecessors establishing a social order built on violence and death threats. Same as it ever was.

As for Kam's description of freedom exchanged for tokens, that's a story everybody knows. Every overworked and underpaid employee around the world might raise an eyebrow at that turn of phrase. It is a motif present in the chronicles of many historic invasions. As Kam describes the process among his ancestors on Molokai it is easy for me to see the trajectory of the change he is describing. The tokens are provided by the new landowning caste. Working for the tokens means you are now working for the new caste. Whoever issues the money, that is who you are really working for. Hey presto, a new class system has appeared with two distinct classes of people. Furthermore, although the process begins with the conquered people *"agreeing"* on the value of the tokens, year by year, decade by decade, the actual value of the tokens is progressively depreciated so that the number tokens needed simply to purchase the same foods, clothes and homes as before, will continue escalating, putting the people under ever increasing pressure as they seek to maintain their quality of life, and although occasional adjustments may be made, the general trajectory is always the same over the long run.

The process of annexing and dividing the land achieves the same result. Before the coming of the Anunu the people lived on the land, farming and thriving on the fruits of the land. After annexation the land is now owned by the Anunu and distributed among the colonizers and those in their favour. This alteration changes the status of the general population. Before the change they were peasants living freely on the land. Now that the lands are owned, the people have become tenants living on someone else's land. The landowners and their overlords will now dispossess the tenants of a portion of their income through the administration of rents, taxes and tithes. If a tenant gains enough tokens to purchase a property, then new taxes and duties can be applied to ensure that one way or another the colonizers can continue to derive a steady upward flow of wealth, while the general population exists in a perpetual state of debt towards the new elite.

With the enforcement of new patterns of ownership and land division, the Anunu effected a simple stratification of society, a hierarchy of a ruling class and working class, with an intermediate stratum of public servants to glue everything together and make it work. In this way the division of land and the creation of new elites, the trickle-up of wealth and the overpowering of the people at large were achieved all of a piece, as the new Anunu overlords installed themselves.

The process that Kam has graphically described to me is analogous to coups and cultural invasions at other times and in other places. For instance, in England, where I grew up, from the 1600's to the 1900's the Enclosure Acts privatized forty-five per cent of the remaining common lands. By displacing the peasant population, these changes drove forward the urbanization of the general population and allowed for better monetization of its agriculture to benefit the new cadre of very wealthy private landowners. This process was then systematized and exported overseas through the agency of the British empire. Through systems of deeds for land, licenses for farming, monopolies for manufacturing and trade, and stamp duty for the acquisition of land and other property, everything the invading Anunu did in Hawaii Britain repeated in its colonies around the world. As recently as the long reign of the late Queen Elizabeth II the same exercises were repeated with ever-increasing violence in the British seizure of Kenya. Similar methods were applied by the Spanish in their fifteenth and sixteenth century invasions of Central and South America. In the Bible, the book of Deuteronomy reports that the same methods were employed as the invading Elohim took possession of the ancient Middle East. Against that international canvas, it isn't hard to visualize Kam's account of the cultural shift brought by the Mo'o, the Anunu and the Ahumanu.

At one level these devices taught by the Hawaiian Mo'o and the Sumerian Apkallu are simply the kinds of conventions required by any society's transition from subsistence living to the larger scale of settled, high-density living which goes hand in hand with the advent of farming. After all, imagine running a city without laws or without money for instance! On the other hand there is a dark side to the division of a harmonious society into those who make the laws and those who keep the laws; those who control the money and those who now have to earn money; those who imprison others and those vulnerable to imprisonment; those who exact taxes and those who must pay them; those who own the land and those who have to part with their wealth just to be able to live on what was once common land. I don't want to overstate it. It isn't that the Anunu or the Apkallu created a dystopian nightmare, rather that they bequeathed to us a matrix of means for managing dense populations in ways that are most convenient for those at the top of the civic tree. In the end it we have to see it as a story of many layers, marked by light and shade.

The same light and shade can be found in the Sumerian narratives of the Sky People, Enlil and Enki, as they conflicted with one another over how best to manage project humanity. Their issues echo in the Biblical accounts of the *El-Ba'adat* (The Council of Power) and in the narrative of Yahweh and the Serpent in the Eden story of Genesis 3. The Mayan-Kiche account of *Los Progenitores* and *Ququmatz*, the Nigerian account of Abassi and Atai and the ancient Greek myth of Zeus and Prometheus, all repeat the same essential conflict over the management of human populations. None of the governing beings in these accounts are human, and so we should probably not be surprised that convenience for the overlords rather than solidarity with humanity appears to be the driver in their pattern of decision-making. And, as Kam reminds me, the ancient invaders, the Mo'o and Ahumanu were definitely not human. Far from it. They were dragons.

Dragon narratives exist all around the world. The work of French mythologist, Dr. Julien d'Huy of the Sorbonne reveals that wherever human beings have lived and recited stories or drawn pictures on stone, tales of dragons have been told. Curiously, the word *serpent* in the English language can refer both to a snake and to a mythical creature with four legs, wings and a fire-breathing snout, something we would all recognize as a dragon. *Serpent* as a crossover word invoking these two quite different kinds of creature has close linguistic equivalents all around the world, which is also curious. In Lesotho, for instance, we can find rock paintings of a creature known as a *serpent* but which is heavy-bodied and four legged. It is emphatically not a snake.

Following the line of d'Huy's research, when the thematic morphology of the world's dragon stories is mapped against what we think we know about humanity's ancient migrations, two distinct families of dragon narratives emerge: an ancient canon, and an even more ancient canon. For an example of the earlier family of dragon narratives, let me take you to ancient China to hear a story which goes like this:

A long time ago all the land of the Earth was barren. It was no more than a wasteland with an ocean to the east. One day, four dragons came flying over the waters. The dragons could see human beings on the land and realized that the people were starving because the land was infertile. Taking pity on the hungry humans, the four dragons flew to the palace in the sky and asked the emperor to arrange for rainfall to make the land fertile. Grudgingly, the emperor agreed and told the dragons that he would make it rain – tomorrow. And so the dragons, trusting the emperor's word, left and returned to Earth.

When tomorrow came, no rain appeared. It was clear to the dragons that the emperor in his palace in the sky did not really care about the human beings at all. So the four dragons decided to break rank and help the humans themselves. Scooping up water from the ocean, the dragons flew over the land and sprinkled it like rain over the barren ground to make it fertile. The human beings rejoiced because finally they had food. So it was that they began to grow and prosper until finally they became a great civilization.

When the emperor learned what the dragons had done, he banished them from the heavens, exiled them to the Earth's surface, and buried each dragon under a separate mountain. However there was still so much water in the dragons' mouths that it began to flow from the bottom of the mountains into the lands to become the four great rivers of China.

This fascinating story is not unique. Stories very similar to the Chinese narrative can also be found in the *Nagas* of Indian mythology, the *Ryu* of Japanese folklore, the *Imoogi* in Korea, the *Horned Serpents* of Native American narratives, the *Barong* of Bali, the *Rainbow Serpent* of Australia, and in the *Rain Snake* myth of Lesotho and other parts of southern Africa. All make the same connections as the Chinese narrative, identifying rain, rivers, agriculture and fertility with a prehistoric intervention from dragon-like beings. D'Huy argues that what we know of humanity's ancient migrations pushes the timeline for these stories far back into the deep past of humanity's great diaspora. Because of the universality of these correlating themes we can date this first wave of dragon story to between 100,000 and 175,000 years ago. Given that the widely accepted timeline for homo sapiens as a species is 200,000 years, 100,000 to 175,000 years is quite a significant age.

My background in Biblical hermeneutics makes it easy for me to recognize significant details in the Chinese dragon story which belong in Biblical narratives too. Many of its elements are present in the creation stories of Genesis:

- A world with one ocean.
- A devastated land, populated by human beings.
- The need for land reclamation and rehabilitation
- Observation by powerful beings, carried in a *ruach* or *wind-maker*, hovering hawk-like over the waters.
- The separation of waters, ocean water from freshwater.
- The fertilization of the land for production of plants and support of animals.
- Human prosperity
- A draconian agent breaking rank to improve conditions for the human beings.
- The banishment of the draconian helper of humanity to exile on Earth.

The Genesis story is, in turn, a summary form of the stories told in ancient Sumer, notably the Enuma Elish and the Epic of Gilgamesh. They too resound the same essential notes present in the Chinese dragon story:

- A world of one ocean
- The separation of the waters, ocean water from freshwater
- Separation of waters;
    - not by four dragons (the Chinese story)
    - or one wind (the Hebrew story)
    - but by four winds
- The compassionless attitude of the superior being governing this region of space (*ie* the emperor in the *'palace'* in the sky)
- The advanced being on earth breaking rank to improve conditions for humanity.
- The demotion and exile of the helper of humanity

A number of these details repeat in the ancestral stories of Nigeria, southern Benin and Cameroon, as told by the Yoruba and Edo people. Their story too begins with a flooded world, terraformed by advanced beings, known as *Osanobua* and his sons, noting the phonic similarity of *Osano (tr. ocean)* and *Oannes*.

Similarly, Filipinos carry the indigenous story of a water-world, terraformed by winds created by the Tagalog, the giant hawk, hovering over the waters.

The ancient Greeks also spoke of the arrival of advanced beings on a chaotic Earth, whom they credited with the terraformation and finessing of the planet's ecology. They called this kind of being the *demiourgos* or *craftsman.*

To be clear, these mythologies from Nigeria, Benin, Cameroon, the Philippines and ancient Greece are not presented as dragon stories. However, they repeat the schema of non-human others arriving in a time of chaos and existential threat to help our ancestors. They are essentially stories of light.

The second wave of dragon narratives emerges at a familiar node in the human story. D'Huy dates the advent of this second wave of dragon narrative at 10,000 years ago. As we saw before, this coincides with the tail end of the Younger Dryas Cold Period, the most recent ice age – another moment in which Earth's human population was compressed geographically, a moment of existential threat for homo sapiens as a viable presence on planet Earth. It is the time associated with Karaca Dag and humanity's great leap forward into agronomy. It is also the moment of external assistance which Berossus associates with the appearance of the *"extraordinary monster"* Oannes, and his peers the Apkallu. What is different about this second wave of dragon mythology is that is marked by dragons who:

- Rule as the non-human governors of humanity.
- Impose their own laws enforced with a threat of violence.
- Exhale fumes which can ignite as fire.
- Are armour-plated, making them virtually unassailable and almost impossible to kill.
- Demand a constant tribute of cattle, sheep, gold and virgin girls.
- Have a horror of female menstruation – hence the continual demand for young girls in particular.

This more recent family of stories contains a sub-canon in which the dragons are finally succeeded by human leaders. In one scenario, the dragon is finally beheaded by a prince. The removal of the head was the only certain measure of the death of a dragon. Two examples of the prince versus dragon scenario would be

Daniel in the Biblical story of *Bel and the Dragon*, and George of *George and the Dragon* fame.

Another pattern within this later sub-canon is one in which the elders of the afflicted community move beyond a state of terror and paralysis to one where fear finally loses its currency. When this moment comes, when enough really has been enough, the elders gather and reason together that if they all act in solidarity, they will have the power to defeat the dragon or, at best, simply inform it that its reign of terror has suffered from the law of diminishing marginal returns and that consequently the humans will no longer fear it or serve it. In those cases the dragon, realizing that its purchase over the humans is gone, retreats, slinking away to the mountains where it retires in lonely isolation. The early Biblical narrative of Yahweh fits precisely within the latter framework of this 10,000-year-old wave of dragon story, fulfilling every single one of the points I have just itemized.

However the story of Yahweh's dismissal by the elders of Israel carries a critical post-script, which can be found in the annals of the Kings of Israel and Judah. Because, for many generations, the human monarchs found they still had to operate with Yahweh metaphorically breathing down their necks, pulling strings from behind closed doors, still defining public policy and insisting on all the usual draconian tribute from the spoils of war – wars of Yahweh's choosing. In this way, the Yahweh dragon narrative takes the 10,000-year-old story even further, thereby offering a Biblical lesson in the persistence of non-human powers and the dangers of covert governance.

The fact that this more politically layered iteration dates from the same period as the Oannes intervention makes me wonder. With all this already in my mind as I listen to Kam's explanations of the Mo'o and Ahumanu, many bells are ringing. When he tells me how the Mo'o stole the lands of Hawaii, enforced laws backed with fiery violence, dispossessed the people, imposed a money-system, and then controlled the money from on high, all the while devaluing it, I cannot hear his account in isolation. The cross-cultural correlations force me to take it more seriously.

Furthermore, when I consider the Mo'olele alongside the Bible's warnings about the covert persistence of non-human powers, I have to wonder how many of our familiar conventions of orderly society might really be the entrenched patterns of an ancient invasion – an invasion whose memory is burned indelibly into our collective psyche. In the present day, might we be simply repeating all the patterns our ancestors learned from cold-blooded others who had the management of large human populations well and truly under their belts? My background in Biblical hermeneutics makes it impossible for me to brush these narrative correlations off as mere coincidence. Too much detail repeats for that to be possible.

Moreover, the Molokai memory of Anunu and the Hebrew memory of Elohim both describe the same cultural shift. One indication of this shift can be seen in the evolving modus operandi of the Hebrew people in the Biblical story. Early in the book of Genesis we find Abraham and his extended household peacefully co-occupying the land on the basis of a matrix of peace treaties and accords. With the exception of one tribal neighbour, Abraham always succeeds in finding a win-win to achieve a harmonious way of life. However, by the time we meet Abraham's descendant Moses in a later generation, something has radically re-ordered human society. As we will see in the next chapter, the ancient Hebrew scribes did not hold back in telling us exactly what that *"something"* was. It was a full-blown alien invasion.

CHAPTER SIX

Invasions and Body-Snatchers

Of course, nobody mentioned alien invasions at theological college. Through all the years of my formal theological training, this way of reading the Biblical narratives was never suggested nor had I ever heard it in any of the churches I had attended. It was only later in my career, through a fifteen-year period of providing in-service training to pastors, that I began to connect these dots. My subjects were *The History of Christian Thought* and *Hermeneutics*. Through these disciplines I began to see the extraterrestrial and invasion stories for what they are. Simply by applying the disciplines of source analysis, form analysis and linguistic analysis, which I was teaching to my pastors, the data gradually mounted, inexorably moving me away from any possibility of reading the elohim and Yahweh texts of the Bible in a fundamentalist way as God-stories.

I am far from the first person to make this journey. In the 1870s when Assyriologist George Smith first translated the Mesopotamian cuneiform tablets into English, the scholars who read his work were quickly able to recognize the Bible's dependence on the stories previously curated by the ancient Sumerian, Babylonian, Akkadian and Assyrian cultures. These were not stories about God, nor were they in any way religious texts. They were stories of paleocontact.

The fact that *Elohim* is the Bible's word for the Mesopotamian *Anunna, Anunnaki* or *Sky People* has now been known by Christian academics for one and a half centuries, yet somehow this information, so vital to understanding how our ancestors viewed their place in the cosmos, has still not managed to filter down to the faithful millions who attend Christian churches from week to week. In my case, it was only the good fortune of an ultimate frisbee injury, and an enforced period of convalescence in between pastorates, which afforded me the opportunity to revisit those source critical discoveries and begin drilling down into the implications of them. So it was an injury with a silver lining. Not every pastor is so lucky.

As I pursued these connections and read the translated cuneiforms for myself, it came as something of a shock to realize that so many of the stories Christianity has presented as God-stories are really a far older fusion of narratives concerning non-human visitors in the deep past whose arrival from outer space was remembered by our ancestors around the world as something of a mixed blessing. The Bible recalls the arrival on planet Earth of the Powerful Ones (*elohim*) in dramatic terms. These powerful beings are known by various names throughout the Hebrew scriptures: *The Powerful One (El) The Powerful One the Prince (Beelzebul) The Powerful one of Ekron (El of Ekron) The Prince of Persia (Wesar malkut Paras), The Powerful one of the Philistines (Dagon) Chemosh, Milkom, Asherah, El Elyon (The Powerful One Higher than the Others) Yahweh* (or *YHWH* tr. unknown) and *The Powerful One the Destroyer (El Shaddai.)*

The collective name the ancient Hebrew writers chose for this incursion of new overlords hammers home the nature of their arrival. They were the *Seba Hassamyim* – the *Sky Armies*. Once established as our governors, the new arrivals became known as the *El-Ba'adat* – the *Council of Power*. To evoke their arrival with the militarized language of *Sky Armies* gives us an idea of the array of craft and firepower which confronted our ancestors. This was the threat now hanging over human society should anyone think for a moment to put up any kind of resistance.

Among this technological array our ancestors recalled the *ruach* – the flying thing which could create vortices of wind so powerful as to drive back floodwaters from inundated land. (Genesis 1.) In the book of Ezekiel the writer uses another name for this craft, *kavod*, and along with the writer of Exodus, describes its modalities of launching and landing, recalling fire and smoke. Ezekiel goes further and depicts what appears to be a drone-like capsule, describing it from the perspective of a passenger.

The book of Isaiah repeats the same threat of violence undergirding the people's compliance to their new overlords. The writer generously reminds later generations of readers of the kinds of consequences on the menu should a group of people ever dare to resist their assigned elohim's right to rule. He writes:

"[The people] refused to walk in his ways nor were they obedient to his laws. So [YHWH] poured upon [his own people] fury from his nose [ap] his great power in battle and set all the people on fire, burning them alive until there was nothing left of them." (Isaiah 42)

Just as there was no love from the elohim towards their allotted humans, there was not a lot of love among the elohim either. Each people group was jealously guarded by their respective elohim. This jealousy is enshrined in the words of the Ten Commandments of YHWH in which the elohim by that name tells his people, *"You will put no other elohim before me. You must not serve them or bow down to them – or even depict them. For your powerful one, YHWH, whose name is jealous, is a jealous powerful one."* (Exodus 20 & 34.)

In essence this insistence on exclusive loyalty to a senior governing power is no different to the way we segregate lands and people groups in the modern world. For instance, when I became a citizen of Australia, I was required to renounce the claim of any other national power over me – other than that of the crowned head of state of Australia, meaning the senior member of the Windsor family in England, who already had dibs on me. However I had to pledge that in an international conflict my loyalty would be to Australia and to no-one else – which is fair enough. So we can view Yahweh's commands to exclusive fealty with that as a frame of reference But whereas in Australia, as a citizen I am entitled, if I want, to have on my bookshelf a picture of myself having lunch with, say the King of Denmark, without any undue consequence, this would have been punishable by death under the strict regime of an overlord like Yahweh.

The book of Ezekiel tells us that Yahweh extended his *"do not depict them"* law to the artwork in the Jerusalem Temple in which Hebrew artists had created carvings and reliefs, to commemorate the invading forces of the Seba Hassamayim as they arrived on the planet in the deep past. The presence of these works among the art installations of the Jerusalem Temple reminds us that the worldview of the Hebrew kings was not originally a picture of Yahwist monotheism. Indeed, the Jerusalem Temple's originator, King Solomon, was a person who embraced the memory of the plural Seba Hassamayim, Yahweh, Molech, Chemosh, Milkom, Asherah and other members of that cosmic cohort. Furthermore, Solomon accommodated and employed priests to service numerous cultural centers dedicated to maintaining the memory of these advanced beings. Solomon's worldview was a canon of paleocontact.

In my book *The Eden Conspiracy* I track the alteration of the narrative through the evolution of Judaism and its Bible, showing how a sequence of Yahwist kings and high priests sought to purge Judaism of its extraterrestrial memories. The book of Ezekiel represents a purge initiated by non-human agents and enforced by advanced, non-human technology. In that book the writer describes a close encounter with a being he describes as a *"human-like lifeform."* The being introduces Ezekiel first to advanced airborne technology and then to advanced weaponry, showing how such equipment can be used for ethnic cleansing. And just in case Ezekiel should be left in any doubt, the lifeform then demonstrates the speed and ease with which mass executions could be achieved on, what was for the iron-age witness recounting it, an unimaginable scale. The lifeform shows him how any family not willing to make a public and convincing show of grief at the presence of paleocontact art in the temple must be put to death using the *keli mashetow* (tr. *The Destroying Device*) or the *keli mappasow* (tr. *The Disintegrating Device.*) To his horror, Ezekiel saw that it took no more than six individuals, equipped with either device, to ethnically cleanse an entire district. This was technology like nothing Ezekiel had ever seen.

Let me just lay out some statistics here to give a sense of scale, when understanding the power of Yahweh's killing technology:

- The assault on the World Trade Centre in New York in 2001 killed 2,996 people.
- D-Day in World War II was responsible for the deaths of 10,400 infantry.
- The Blitz of London killed 40,000-43,000 Britons.
- In one single assault, recalled in the book of II Kings, Yahweh disintegrated 70,000 of his own people.

Because of translation choices made in the final redaction of the Hebrew scriptures in the C6thBCE, these stories have been read down the ages through the lens of idolatry and divine punishment. Absent of that lens and interpreting from the root meanings of the key words in these texts, what emerges is a story of invasion, colonization, narrative control, and martial law. Every one of those elements is now emerging in Kam's telling of the Mo'olele. Only, when Kam describes the prehistoric invasion of what was to become Hawaii, he tells the story in the language of dragons.

If I had sat at Kam's feet only five years ago, his dragon-language would have prejudiced me to listen to him, all the while mentally consigning his stories to a bin labelled *"fable and fiction."* Now that I have listened to more ancestral and indigenous stories, I realise that there are many layers to the world's dragon narratives and my respect has been elevated by the international repetition of surprising details of realpolitik in this ancient canon, for all the reasons we saw in the last chapter.

On the island of Molokai, for all the drama of their violent arrival, it was what the Mo'o and the Ahumanu did next which really grips Kam as he continues to unfold their story. Kam's reportage doesn't focus so much on airborne armies, advanced craft and armaments. His stories are not of bombardment, massacres, or warfare. Rather they shine a light on the cultural shifts which followed the establishment of the new uber-government – the loss of land rights, the loss of freedom, and the enslavement of the people through a money system instituted and controlled by the new ruling elites.

Kam's themes of annexation of land, its division and allotment to new powers, who then exert new laws, are themes which echo in the stories of the Bible. The moment in the book Deuteronomy in which lands in the Middle East are sub-divided into separate territories and parceled out to the elohim by the new overlord, El Elyon, gives us eyes on what is clearly the immediate aftermath of an invasion. A new order and a new form of governance have arrived. Beings who were not previously part of the picture are now calling the shots. Any previous patterns of land ownership, or treaties of co-occupation have been deleted and replaced with this new order of top-down management, in the face of which any grassroots resistance will prove futile. The overarching picture of separate human colonies, governed over by competing non-human governors then plays out through the books of the Bible which follow.

Of course, this a scheme which spreads further than Hawaii and the pages of the Bible. The arrival of the Norse *"gods"* or Aesir in northern Europe showed itself in the sagas as an annexation of previously populated lands by Odin and his cohort, who then distributed those lands to his sons as if they were his own possession. Clearly, native title meant nothing to the Aesir. In the same way the great conqueror Haldi took the lands of ancient Armenia to govern its people as his own. Similarly, in Africa, the Luo and Yoruba people describe the arrival of Osanobua who rehabilitates lands in Nigeria, southern Benin and Cameroon, and then redistributes those lands among his sons, who in turn parcel them out among the Ogiso. So when the Mo'o arrived in ancient Hawaii they were following a well-worn path when they annexed the land and redistributed it in defiance of any patterns of ownership prior to their arrival.

In Deuteronomy 32, El Elyon follows the exact same pattern but adds a further stressor to the scenario by apportioning to Yahweh a people-group with no land for which the writer gives no reason. This action creates an artificial scarcity which immediately primes the elohim for conflict and competition. For a people-group to be invaded and dispossessed of their homeland, and thereby rendered landless and without citizenship, is always a profound injustice and can only lead to warring and suffering.

Accordingly, when we first meet Yahweh in his encounter with Moses in the book of Exodus, his first task is to extract his people-group from lands governed by another elohim, Akhekh of Egypt. Having successfully commandeered his allotted people group, Yahweh then imposes laws against which his people will now be policed by his representatives. In the stories of the Mo'olele and the elohim of the Bible we see the same gradual disempowerment of the people, beginning with the annexation of the land.

With all the devastation of the 2023 fires on Maui, I know that for my friends Kam and Kalea these ancient stories hold a deep and painful resonance and I am very aware that right now they are grieving for friends who have lost homes and are mourning others who have lost their lives.

"Of course the newer, wealthier properties have fared far better because they are built to code, with concrete and fire-resistant materials. It is the older traditional homes that have gone up in flames, and families who have farmed their lands for generations are left with nothing."

"We have fires here and we have experienced losses before, but it makes me angry when I hear from my friends on Maui. Some have lost their homes and some have lost their loved ones too. Yet while they are grieving, within days of the fires coming through, they are receiving phone calls from agents. These agents represent the big landowners, people who already have hotels and lands and more property than they know what to do with, and the agents are asking, 'Would you like to sell your property to us?' Can you believe that? These agents are approaching my friends at their lowest point to ask them that!"

"At the very least it shows a total lack of human compassion. At the worst I would call it a landgrab. That's what it feels like, to be honest. The fires are being exploited to enable another landgrab for the American corporations. More and more Hawaiian people are being pushed off their ancestral lands. Before the fires, American billionaires were already coming in and using Quiet Title laws to acquire land that was already owned. Somehow our courts have been favouring the new privatization of land over the

traditional models of land ownership by the Kanaka Maoli, the traditional owners."

"That's why today Hawaii is the most expensive state to live in the U.S.A. It is why today you can find traditional owners of the land, living in cars and tents."

"In 1993, when America apologized for the invasion, many of us hoped that things would change for the better. But look at what is happening now. It is another overthrow of our people and our land. It makes me very sad. It is the way of the Mo'o. They marked their arrival with destruction by fire, and then seized and privatized the land. It is history repeating itself."

I can feel Kam and Kalea's sorrow at what has happened in 2023, and I know they in turn are burdened with their friends' grief and anger at how this tragedy has been managed by the authorities. There does seem to be a bigger picture here and it does feel like a piece of history repeating.

You and I are traversing some dark territory in this chapter, and I want to thank you for keeping me company this far. These are stretching concepts for many reasons, and I promise you that things will lighten up in a moment, when we will pause for a coffee with my pastor-friend, Will. He has a bee in his bonnet about what's happening in Washington and I know he wants to talk it out. But, before our coffee break, we must touch on another dark layer of Kam's ancestral account of draconian invasion. Were it not for my background as an Archdeacon for the Anglican Church in Australia, it is a layer to which I might not have given due attention. However, informed by that experience Kam's story raises some bright red flags the moment I hear it.

"Our whole world was turned upside down to benefit the Mo'o. The Ahumanu (the bird-people) took our children and terrorized

them. They extracted something from their blood which gave energy to the Ahumanu."

"Bird-people" occur in the visual canon of cultures all around the world. They appear on Easter Island, in Mesopotamian carvings, Egyptian Hieroglyphs, and Mayan reliefs. Biblical narratives of Yahweh reference that being's pinions or flight-feathers. Though when Kam tells the story of ancient bird-people predating upon human children for something in their blood, my imagination goes not to these ancient carvings and Biblical texts, but to more recent cultural tellings. I am thinking for instance of Jim Henson's *The Dark Crystal*, a movie which relates the story of an ancient world in which callous *bird-people* rejuvenate themselves through extracting the *"essence"* of terrorized, innocent, child-like gelflings. A coincidence? Did Jim Henson know the stories of the Ahumanu? The same motif is there too in the movie *Monsters Inc* in which the Monsters extract and bottle the *"terror"* of children being tormented by nightmares. Even *The Simpsons* gets in on the act in the episode in which the desiccated old tyrant, Mr. Burns, has to be rejuvenated on a regular basis with the blood of a young Bart Simpson.

Where is this current of parallel narratives flowing from? What have these scriptwriters been reading or hearing? I may be naïve, but when I watch fiction like this I don't for one minute expect to be imbibing anything other than pure fun and fantasy. At least that was the case until my years serving on the Senior Leadership Team of an Anglican Diocese. My service at this senior level of ministry coincided with the period of a Royal Commission which probed into institutional responses to child sexual abuse, abuse which I soon learned is often organized and facilitated at elite social levels. Unfortunately, the remit of the Royal Commission was pointedly *"institutional responses"* to abuse not the *"institutional organization"* of it.

I remember vividly the feeling of dismay in the room as one of our senior leaders in professional standards quietly explained that the organization of abuse by Christian institutions would not be scrutinized by the commission, simply because it would implicate too many powerful people. Suffice it to say that the things I learned in that chapter of my career mean that I can no longer watch the *Dark Crystal*, or *Monsters Inc* or even that episode of *The Simpsons* as innocently as I once did. So when Kam tells me the story of the Ahumanu taking the ancestors' children, tormenting them and extracting their terror, I am not able to make the lazy assumption that I am listening to fiction or fantasy.

Similarly, the ancient Mayans associated a pattern of child-sacrifice with the approval of their feathered serpents. Why would that be? What exactly was to be gained from such brutality and terrorization? Was it pure power play over the citizenry by the rulers? Or is *The Dark Crystal* closer to the mark?

Biblical writers also retain a memory of ancient patterns of child terrorization and murder, known as *"moloch sacrifices."* These inhuman demands were a device by which the governing elohim would display total control over their human subjects. If an elohim could get a human being to do such a thing, especially to a first-born son, it was a show of total control. *"If you can get human beings to do that, then you have completely broken them. There onwards you will find you can then get them to do anything."*

This is the ugly background to Abraham's infamous near sacrifice of his firstborn son, Isaac on Mount Moriah. It is also present in Hawaiian memory of the Ahumanu and their successors. Why would cultures as disparate as the ancient Hebrews, Mayans and Hawaiians all, separately and independently create such a similar and inhuman paradigm? Another coincidence?

99

Recognizing the correlations in the past and present, I have to take it as entirely possible that in Kam's story of the *Ahumanu* I am hearing the memory of a collective cultural trauma. Incidentally, *Ahumanu* is another word which catches my attention. It looks suspiciously like Latin word parts which would imply beings who are *"inhuman"* or *"against human beings."* Judging by Kam's stories of them, they certainly fit that bill. Neither were Kam's adult ancestors spared unwelcome interference by their inhuman overlords. As the evening wears on Kam relates more stories of the Mo'ohe'e.

"When the Mo'ohe'e came they came from beneath the ocean. They would come onto the land and seize people, young men and women, and take them back with them into the sea."

"Under the cover of the deep ocean the Mo'ohe'e lived in underwater cities, and that is where they would take our ancestors and use them to make hybrid people who were part human and part Mo'ohe'e. They wanted to change themselves to look more like us, so that they could move among us and live more on the surface. *When our relatives were taken, we never saw them again."*

A few short years ago I had no frame of reference by which to handle a claim of this kind. This changed for me when I took my life in my hands and published *Escaping from Eden,* my first book of paleocontact. From that moment on people began reaching out to me, with stories from their own families which would strongly suggest that Mo'ohe'e abductions are far from a thing of the past. What was even more of a shock to me was to discover this same experience in the story of my own extended family.

On my maternal grandfather's side of the family, my Welsh ancestors told the Mo'ohe'e story in the narrative of *Tilwith Teg.* It repeats the Mo'ohe'e narrative in every note except that in the

Welsh telling the underwater hybridizers give the appearance of being human though ultimately, they prove themselves to be something else. More than human, the Welsh people-stealers are described as being irresistibly beautiful. They take the young men away to their underwater bases not by force but by enticing them in every which way. On the other side of my family are the traditions of Ghana, West Africa. In fact it was my Ghanaian parents in law, Kofi and Patience, who first introduced me to the Ghanaian iteration of this story – the Mami Wata tradition.

Again, according to the Ghanaian tradition, victims are not taken by force, rather they are enticed with promises of adventure, health, wealth and heightened abilities. Oddly, both the Ghanaian and the Welsh stories include the unlikely aspect of abductees being returned a few years after their disappearance, having been used for hybridization and now being surplus to requirements. My next surprise was to learn of a close family connection with a family from Anloga, in the Keta district of Ghana's Volta region, who had experienced this phenomenon within my lifetime.

These discoveries sent me on a global quest to hear this same phenomenon described in Kenya by the Luo and Maharani people. Filipino brothers and sisters taught me about the *dili ingo-nato* and the *engkantos* and the *duwende*. In Trinidad and Tobago the story is of the *douen*, and in Belize those who abduct are the *duhende*. In Lae, Papua New Guinea stories are told of the *Hena-sekato*. In Cuba and the Caribbean you will hear the stories of *Yemoia*, and now Kam is telling me the same story and using a word with curiously similar phonetic features. A shared ancestral form of Mo'ohe'e and Yemoia would be something like *mm-oe-y-oe*.

Then, just a few weeks ago, as if to remind me not to disregard the abduction aspect of Kam's story I received a phone call from another member of my extended family in Ghana.

"Paul, we were talking about your book today. My sister and I were remembering how Mami Wata was known in our district. We were never taught about it in school but in our families, we knew about these abductions. As we talked together my sister reminded me of an auntie of ours. She said, 'Don't you remember, Kuor, Auntie Grace was gone for three years and nobody knew where she was?'"

"'When she came back, she told us it was the Mami Wata people who had taken her and that they had kept her in a city under the ocean. She told us she had children while she was with them, and when she came back her doctor confirmed that she really had given birth to children.'"

It is amazing, the stories in our families which go untold!

"Paul, after she came back to us our Auntie was different. Even though they had kept her by force she missed being with the Mami Wata people and she said they were very beautiful people. Every morning she would go down to the beach and stare at the water. Our family all laughed at her and said she was 'worshiping idols in the water.' I don't think she was. I think she was grieving."

"But even our relatives who laughed at Auntie knew that something real had happened, and it was something very strange. Because in the time that she was away somehow her eyelids had been altered and her eyes would often stream with tears. And my sister was right. Even though our relatives did not want to believe what our auntie was telling us, we knew something had happened to her that we could not explain."

Since publishing *Escaping from Eden*, I have lost count of the number of people who have contacted me, spoken to me, or met with me, who have told me, hesitantly, nervously, that they too carry a memory of abduction. As with any pastoral conversation one listens without prejudice and respects the person's subjective interpretation. At the same time one has to filter, assess, consider whether some pathology may be present, yet all the while keeping an open mind as to what might be the reality behind the memory. However over time, it becomes clear that these experiences are not shared as if to boast or attract attention but are instead held very close because of the trauma and embarrassment associated with these experiences. Nick Pope tells me that this kind of data is what researchers within the British Ministry of Defense and the U.S. Department of Defense obliquely refer to as *"human interface."* It is such a dry and innocuous-sounding phrase, but this is what they are talking about.

Both Kam and Grace speak of underwater bases or *"cities."* Indeed the motif of underwater bases repeats throughout the global testimony of abduction and hybridization. This is relevant to today's conversations in Washington DC and Arlington Virginia because of videos of UAP engagements by US Defense in 2004, leaked by former Assistant Secretary of Defense Chris Mellon in 2017, and authenticated by the Pentagon in 2019. These videos include footage of anomalous craft travelling at incredible speeds through the air and into the sea. The pilots tracking the craft can be heard commenting on this trans-medium ability. To them, the movement of these craft implied the existence of covert underwater bases. The language invoked by the Pentagon today of *All Domain Anomaly* or *Trans-Medium Anomaly* is an acknowledgment of anomalous craft which can operate seamlessly in space, in the air and under the water.

Another pattern in the world's hybridization stories is that the non-human hybridizers are generally not presented as new arrivals or invaders Many narratives acknowledge the hybridizers as having been present underneath our great lakes and oceans for a very long time. It is a picture of longstanding co-occupation of our planet by non-human others.

For these reasons, I note David Grusch's use of the words *non-human* and his careful avoidance of the words *alien* or *extraterrestrial*. Bear in mind, this is the vocabulary respectively approved or proscribed by DOPSR. When quizzed, David Grusch says he doesn't want to *"narrow down"* potential explanations. *Alien* implies very different to us. *Extraterrestrial* implies they don't live here. However *Mami Wata, Engkantos, Henasekato,* and *Yemoia* stories, the Luo and Maharani traditions, all state that the hybridizers look a lot like us, with the Filipino narrative noting only an odd skin colour and the lack of a philtrum on the top lip to distinguish them from regular human beings. Our ancestral stories of abduction identify non-humans but don't pin them as aliens or extraterrestrials. I wonder if this is what David Grusch may be hinting at.

Kam's explanation of the Mo'ohe'e differs in two respects to the general scheme of these ancient stories. Firstly the Mo'ohe'e took his ancestors by force, and secondly, they were not co-occupiers. They invaded. When I mention this difference to Kam, he is quick to remind me, *"Paul, remember I am from one of the old families. Our memories go back a long, long time. Before they co-occupied our world the Mo'ohe'e first had to arrive. Before they were 'co-occupiers' they were invaders."*

Because of the global context of Kam's story I am ready to listen to his explanation of the Mo'ohe'e with genuine respect and I am writing as fast as I can to keep up with everything Kam is telling me.

"Kam, if I can just remember these names, and these key words, I think I will be able to remember the whole story."

"I know," says Kam. *"That is how it works. That is why our stories are centered on genealogies. It is also why, for nearly one hundred years, the U.S. government made it illegal to speak these words or pronounce these names on Hawaiian soil. Of course if you cannot speak the names, then the story is forgotten. So that is how it was. It was against the law to remember."*

Even before Kam speaks the words I already know. He doesn't have to tell me. I already know the period he is going to identify. I know because the same suppression of indigenous memory was attempted simultaneously in the mainland U.S.A. in Canada and in Australia. These are three separate countries, each operating under their independent, democratically elected government. And yet, by some means, simultaneously, they decided to illegalize indigenous languages, outlaw traditional initiation ceremony, kidnap indigenous children and imprison them in detention centres euphemistically referred to as missions and boarding schools, run by Christian missionary agencies and church denominations. Here the children were forbidden to use their traditional names or speak their traditional languages. The ministers, monks, priests and nuns shaved off the children's hair and terrorized them. And the reason was the same across all three countries; to discontinue indigenous identity, memory, story and ability. Across all three countries the period of these policies was the same, lasting from the 1880s until the 1980s.

In Canada there is a *National Truth and Reconciliation Commission* intent on surfacing the evidence of what its chair Judge Murray Sinclair has called a *"cultural genocide."* The brutality and violence of these policies, coordinated over the same one-hundred-year period indicates how seriously the international uber-government took its mission. By *uber-government* I mean the shadowy powers who sat over and above the long sequence of democratically elected governments in those three separate

105

countries, who came and went through that one-hundred-year period. An authority higher than those three national governments must have set the goal and maintained the policy to eradicate the narratives and any higher cognitive abilities carried by the indigenous peoples of those three countries.

For all these reasons I am not surprised that the period in which Kam could have been prosecuted and imprisoned for telling me everything you have now read in this book, lasted from the beginning of Hawaii's annexation by the USA in 1893 right up until 1983. We have to ask, *"What is the power of these narratives that they were so feared by the U.S. government and the international uber-government? What official narrative did they challenge? What church orthodoxy did they contradict?"*

This morning my mind is still spinning with questions. In 2020 Haim Eshed said that our covert governments have agreed with our *"non-human"* visitors' program of research. *Research*: is that what the Mo'ohe'e or the Ahumanu would have called their abductions I wonder?

What has really resonated with me in my conversations with Kam is that when he speaks of ancient invasions, he doesn't talk about anomalous craft or advanced weaponry. Instead he speaks about shifts and changes in culture. He speaks of a rapid transition from harmony to division and competition; from community to selfishness and greed, from working for one another to slaving for masters, from enjoying the fruits of the earth, to laboring for money. All in all Kam has left me in a pensive state of mind. If these cultural shifts really are the signatures of an invasive non-human elite hijacking the reins of Earth's governments, then.... But no. That's too dark a thought for such a beautiful morning. It's time for that coffee.

CHAPTER SEVEN

Anomalies and Biologics

The Bakery on the Beach - September 2023

"Paul, how can you take anything seriously that comes out of the theatre of American politics? I wouldn't take anything in Washington at face value. I mean, there is no way that hearing on July 26th would have been allowed by the Pentagon unless they had some pre-planned endgame, and I'll bet my bottom dollar the intended outcome has nothing to do with transparency or disclosure. David Grusch is an intelligence officer. He's got no freedom. What he can and can't say has been decided by others. Doesn't that automatically make him a shill? Yet you're taking him at face value. Aren't you smarter than that Paul?"

Will and I are at the bakery on the beach, enjoying what was a quiet coffee, until we got onto the topic of the current process in Washington. Will is on a roll.

"As for any real disclosure, isn't what Grusch says a bit thin? I mean first off, Grusch never says the pilots of the retrieved craft are ETs. He just keeps saying 'non-human biologics.' I mean what is that? He doesn't say. Second off, he doesn't say he has any expertise in UFOs or UAPs. His remit was 'Trans-medium anomaly,' whatever that is. And third off, he says he has never had eyes on any alien technology. So why should I listen to him?"

My friend Will is not an out and out skeptic. He's just cautious. He is not closed to the possibility of other life in the cosmos, nor that our ancestors may have had contact experiences in the deep past. Indeed he's read all my books on paleocontact from *Escaping from Eden* all the way through to *The Eden Conspiracy*. So I owe him a respectful reply.

"Well Will first of all David's remit was Trans-Medium Anomaly. That is Military Intelligence nomenclature for UFO's. Since July 2022 the Pentagon's unit for studying UFOs has become the All-Domain Anomaly Resolution Office or AARO for short. So it's not just focused on things which fly in the sky and which can't be explained. All domain basically means in space, in the air and under the water. That's what 'All Domain' is about. Anomalous craft which can operate in all domains, space, air and underwater is called a 'Trans-Medium' anomaly, and studying those was David Grusch's remit."

"As for David Grusch's use of the term non-human biologics, let's piece that apart. Remember that everything Grusch is telling us publicly has been cleared by DOPSR. He is going right to the edge of what has been officially cleared for public consumption. So where he stops is always significant. Each stop tells us to pay attention because the next detail is what he is not allowed to tell us, OK?!"

"So David Grusch uses the word 'biologics,' a word which means 'living things.' So at the July 26th hearing, Representative Nancy Mace asked the obvious next questions, 'Do we have the bodies?' Answer, yes. The next obvious question is 'Were they or are they alive?" Hey presto, that's another question for the SCIF!"

"In choosing 'non-human' over 'alien,' David Grusch says he wants to keep the picture open. I think that's telling us quite a lot. Firstly because 'alien' implies different to us. Perhaps they're not that different. 'Extraterrestrial' implies they are not from planet Earth. Whereas perhaps they have been part of a covert terrestrial presence for a long time and he doesn't want to rule that out, for some reason."

"Basically, the questions Grusch doesn't want to narrow down are to do with what the non-human biologics look like and where

they are from. Maybe there is more than one answer to those questions.

"And while we're thinking about words, Will, don't miss the ancient stories of 'spirits.' Some of our distant ancestors used that word to describe our visitors because of their ability to ping in and out of our airspace in ways our ancestors couldn't understand. Really, everything we are hearing today around The Program and the UFO phenomenon has an ancient precedent.

"As for eyes-on, yes of course that's what we would all like to hear. But that wasn't David Grusch's job. His remit was to get to the people with eyes-on. It was a job with Congressional and Pentagon authority behind it and which authorized him to get an overview and detail of The Program. Yet despite all that authority behind him he was blocked. So he had to put his hand up and say, 'I am not the obstruction here!' That's what the hearing in July 2023 was about."

"To me it doesn't make sense to dismiss David Grusch as a shill. I don't think there's any logic to that. Who does it advantage to suggest that one arm of government is illegally withholding vital information from another arm of government? It's a story that flatters neither the military intelligence community nor Congress. It makes one look shifty and the other impotent."

I'm not telling him this, but I actually agree with Will that the Inspector General of the Intelligence Community cannot have escalated David Grusch's complaint without some kind of an end game in mind, and it is an endgame which might have nothing to do with greater disclosure of ET contact. As to what that other agenda might be, we will get into that in a later chapter. My goal in this conversation is to encourage Will to keep his feet on the ground.

"Basically, for the time being, it boils down to this: Congress is being kept in the dark and there are certain members who don't like that, and I can quite understand why. They believe their exclusion from this information is unconstitutional. But are they right? And is David Grusch right when he says that his exclusion from information about The Program was illegal?

In Great Britain, where I grew up, intelligence services keep democratically elected parliaments, governments and prime ministers in the dark on principle. This is because the secret services derive their authority not from parliament but from the Crown. It is *"His Majesty's Secret Service"* and this has been the arrangement since the 1500s. For this reason British governments have had a long time to come to terms with their exclusion from the world of international intelligence. Other than on a need-to-know basis, parliament and the prime minister can expect to be told nothing.

In the U.S.A. on the other hand, any appeal to Crown authority was done away with two hundred and fifty years ago. What then is the legal basis for the same powers of secrecy to exclude members of Congress and even the president? Furthermore, what is the legal basis for the withholding of vital information by one American intelligence agency from the next? The phrase, *"a house divided against itself"* comes to mind.

"So, Will, I was talking to Richard Dolan on 5thkind.tv the other day and we got onto the legality question. I asked him what he makes of it and I thought what he said rang true."

In case you're not familiar, Richard Dolan has been a high-profile researcher in the field of Ufology for a long time and is widely recognized as one of the most authoritative commentators in the field. His work is marked by intelligence, good humour and meticulous fact-checking. My work on *5thkind.tv* allows me the

110

privilege from time to time of comparing notes with Richard, and I appreciate his perspective as a fellow author whose feet are well and truly on the ground. This was his take:

"I would say that Grusch is technically correct when he says it's illegal – and he knows the secrets that I don't know. But I do know that the United States Intelligence Community is accountable to the president, and the president, legally, is responsible to the citizens of the United States. We don't have a crown. The people are sovereign. Legally, technically, that's how it's supposed to be. But... it could be a legal illegality! It's illegal but they make it legal! For example, the whole idea of a black budget is not supposed to be legal in the United States, but we still have them. They're called 'Special Access Programs.'"

More about black budgets and special access programs a few pages from now. Meanwhile, I put Richard's explanation to Will, for his take. Will remains unmoved.

"Paul, you do realize that all this clanging of cymbals might lead to nothing more interesting than some grey-suited government attorney standing up and telling us all that, technically, the Pentagon is just doing its job?"

It's possible. However I don't share my friend's skepticism. I actually think something more interesting is in the pipeline. I think what we are seeing is an escalation of insurance against disclosure.

"What I am thinking, Will, is that ufologists used to say that the U.S. government would never come clean about past ET contact because of the political fallout of admitting they had been lying to us. In a way this current piece of theatre could be the government's way of avoiding that kind of fallout and loss of trust, by throwing military intelligence under the bus instead."

111

"Essentially, they are saying, 'Oh no we're not the secret-keepers. It's the Pentagon! They are the ones who have been keeping the public in the dark – and us too. We're as shocked and offended as you are. It really isn't our fault!' Only, frankly, I am not sure that's much of a better look! I just wonder if this kind of play might be a template for disclosures to come."

I am not the only person to think that this is not a good look for the U.S. government. My fellow Australian, the investigative journalist, Ross Coulthart, makes the same point. Speaking on the news channel *Newsnation* he put it squarely and simply.

"Does the government want this to come out in a spill inside the Congress? Or does it want to present as a government that's in charge, that's able to tell the public the truth. It's as simple as that."

I have to agree.

Meanwhile, on Capitol Hill, Senator Marco Rubio has boldly put his shoulder to the plough in an attempt to turn up the heat between Congress and The Pentagon even higher by adding a sixty-four-page amendment to the fiscal 2024 National Defense Authorization Act to accelerate disclosure by empowering Congress in the whole domain of UFO intelligence. His amendment, which successfully evinced bipartisan support, proposed to give Congress far greater power to demand information. These sixty-four pages include UAP disclosure measures, tabled by Senator Chuck Shumer, the Majority Leader of the U.S. Senate.

However, it takes a lot to trump the laws which empower top secrecy in military intelligence. These are the laws that kept the Manhattan Project under wraps. They are the legal matrix which in the time since has empowered the Pentagon's secrecy over The

Program, and successfully so for more than seven decades. That knot of laws and protocols may take a bit of unravelling. I am not confident that Washington can really outgun the Pentagon simply through passing new pieces of legislation. A few chapters from now we will have a clearer idea of where this legislative approach might be headed. My friend Will may not be an out and out skeptic, but he is still not convinced.

"But don't you think it's odd, Paul, that all this noise is about America? There's a whole planet here. Your books are all about ancient narratives in every part of the world. For me it's that international voice that gives what you are saying an ounce of credibility. But when it's just the U.S.A. dealing with 'All Domain Anomalies' doesn't that seem a bit suspect?"

Of course it isn't only the USA. The background to these latest machinations is an international array of voices outing contact which has been going on for decades in the modern era, and in the longer picture, for millennia.

In 2019 in France, the former chief of French Intelligence, Alain Juillet, evidently with government clearance, told the world's press that he was there at the inauguration of AATIP's oversight of *The Program*. AATIP (the *Advanced Aerial Threat Identification Program*) was Lue Elizondo's unit within the Pentagon, and Monsieur Juillet was happy to confirm Lue Elizondo's public statements about its remit of examining materials engineered in space and obtained from UFO retrievals.

In 2021 it was the government of Israel which took a step forward, when it allowed its former chief of space security Professor Brigadier-General Haim Eshed to make a public statement about decades-long and ongoing collaborations with a spectrum of galactic visitors.

In the 2000s, without any Government debunking or action against him, it was the Canadian former Minister of Defense, Paul Hellyer, who was allowed to speak of extraterrestrial contact going back to the 1960's.

2009 saw Vatican City produce senior spokespeople, the Senior Astronomer for the Vatican, Reverend Doctor Guy Consolmagno, alongside Fr Jose Gabriel Funes, the Director of the Vatican Observatory, who briefed the press and told the world to be ready for extraterrestrial contact *"sooner than anyone expects."* Monsignor Corrado Balducci, senior papal advisor on the paranormal, affirmed on the basis of his own research that individuals and groups who report close encounter phenomena are not victims of psychosis or demonic influence. Rather they are experiencing contact with *"a totally other kind of being, one which merits serious study."* This was noteworthy.

In that same decade the Russian government allowed then Prime-Minister Dimitri Medvedev to refer on an open mic to the dossier which is given to each prime minister on their arrival in office, detailing the spectrum of space-faring civilizations with whom we are already in communication. The footage was not seized or censored. Neither was Medvedev distanced or debunked by the president or any other official authority. This departure from the general script of silence or debunking around UFO incidents was also noteworthy.

In that same period, the United Kingdom was busy altering its extradition laws with the United States of America, to prevent the arrest of a British software engineer by the name of Gary McKinnon who had successfully hacked NASA's computers. I say *"hacked,"* but I think it's fair to say that when your desktop's password is *"password,"* it doesn't take a hacker to get into your computer and read your files! The particular files which Gary McKinnon surveilled included images and texts which evidenced exactly the kind of collaboration about which Brigadier-General Eshed was to speak openly more than a decade later. Although parliamentarians in Westminster, by some triple whip, did not discuss the exopolitical implications of the data McKinnon had obtained, the evidence of collaboration was the background to the whole controversy between the two countries. By discussing these matters in parliament, the shadow, at least, of ET contact and collaboration was now on the record of British parliamentary proceedings.

"So no, Will, it isn't all about the U.S.A. What is happening now in Washington is really part of a bigger picture."

I share all this with you today confidently and straightforwardly. Yet a few short years ago I could not have told you any of this – neither the contemporary news around ufology, nor the ancestral consensus around paleocontact. At that time I was in the thick of the world of Christian ministry, coming to terms with the disturbing territory into which my work in hermeneutics had led me. I never saw it coming, but my journey in Biblical translation was about to carry me far away from the cozy world of church-based ministry and win me more self-appointed religious critics in the world of the church than I could ever have imagined. Hence the bloody, medievalist threats I am about to share with you in the next chapter.

CHAPTER EIGHT

Beginnings and Catastrophes

Queensland, Australia - 2024

"It's going to be a bloodbath when Jesus returns...You have no idea what's coming! You will have to forgive my intense disdain for your views, because what you are saying is heretical and blasphemous. You are clearly a freemason, a Bolshevik and a psyop, and it is obvious to me that you are under satanic bondage. You are leading people away from the truth to Satan, and are going to have a long, long time to repent, but it will be too late...Good luck to you!"

Robert is clearly annoyed with me. He is offended because I have had the temerity to question his view that the Bible is *"God's love-story to humanity."* Because I disagree with his *"love-story"* interpretation, apparently, I now deserve a horrific eternity of fiery torture to be meted out by the divine hero of his love story. Credit where it's due though, Robert certainly has a way with words.

It's early on a Monday morning and I am at my desk, catching up with a weekend of emails, direct messages and comments on the Paul Wallis channel. My fiery correspondent is one of a number this morning who are eager for me to know just how violently they object to my most recent book, *The Eden Conspiracy*. Robert has not read the book and, apparently, he doesn't need to for him to know just how wrong it is. Thankfully, Robert is one of only a tiny minority among my correspondents. But whether they be fiery or friendly, I always love engaging with my readers, a group which includes many Christian and Jewish academics and other senior church people, most of whom tell me that they are relieved that I am putting taboo questions on the table, questions long overdue for an intelligent review.

After thirty-nine years in church-based ministry and thirty-three this year as an Anglican priest, I am very familiar with the strong, devotional relationship which many Christian believers enjoy with the Bible. Yet as we saw before, a foundational aspect of the Hebrew canon is a record of invasion, land division and the annexation of ancient human societies by *"non-human biologics,"* a vital aspect which is usually missed and seldom preached. Putting forward the case for this interpretation, as I do in my books, on the Paul Wallis channel, and on *5thkind.tv* is, as you can see, quite stretching for some religious believers to do business with, but I always do my best to engage with viewers for whom these thoughts may be confronting.

So, rather than lazily dismiss Robert as an angry nay-sayer, I take my time to craft a polite reply, pointing him to the reasons for my position, reasons to do with understanding the language and history of the Bible. This may not be a very strategic use of my time, as I do get a lot of these messages, but I know personally how hard it can be to break free from religious entrainment and I recognize the sincerity of some as they battle to hold onto beliefs which they have been taught to regard as more important even than life or death.

Seeing the Bible as a book, bound by two hard covers, gives the casual reader the suggestion of a single narrative, running in neat and tidy chronological order. However, as I show in *The Eden Conspiracy*, this simply isn't so. Indeed as any scholar of the Bible can affirm, the Bible is in reality a library of ancient texts, compiled, redacted and re-ordered into their current form sometime in the C6thBCE, and there is a very broad scholarly consensus which would affirm that. Although the 6th century redactors did their best to massage the diversity of texts into a coherent whole, framed as a story of monotheism, the action within the narratives and the root meanings of the key words bleed through the top layer of redaction revealing the indelible outline of what these stories were before they were changed.

In the pages ahead I am going to suggest that among the succession of Hebrew stories, is a sequence of narratives which memorialize several dramatic turning points in the human story as carried in our ancestral memory. In the pages of Genesis I will highlight seven stories of beginnings and seven ancient cataclysms - almost all of which are marked by extraterrestrial contact. Let me show you what I mean.

Seven Beginnings

Genesis 1 – A Reboot of Life on Earth

Genesis 1 tells the story of a chaotic world. Before the atmosphere is cleared to make visible the light of the sun, moon and stars, the planet is already here, but laid waste by floodwaters. The Hebrew phrase is *tohu wa bohu* and it indicates the condition of an environment that has been ecologically devastated.

Powerful beings (*elohim*) now arrive on the scene in a wind-making machine (*ruach*) which hovers over the floodwaters. Having first cleared the soot-black atmosphere the ruach begins to separate freshwater from floodwater, and land from ocean, gradually rehabilitating life on Earth, beginning with the oceans, the plant life, then animal and then human. This story, which reiterates the Sumerian story and echoes in Mayan, Nigerian, Filipino, and Iroquois narratives, is told from a survivor's perspective, with the Chinese dragon version noting the compassion that the flying visitors felt for the starving survivors who are clinging to life on the remaining land. Like so many *"creation"* narratives from around the world, Genesis 1 is really not a story of creation at all but a story of new beginnings.

Genesis 2 – The Advent of Human Males and then Females

Genesis 2 tells a more detailed story of the creation of humans. First the male is engineered from organic matter available on Earth. This is expressed as clay. The Sumerian story also has humans coming from clay. The Mayan story has humans being made of maize. Bizarrely, the female is produced as an after-thought to the male, hardly a natural sequence, but that is how the Genesis story goes and this strange aspect repeats in the Sumerian account of Enkidu, the primitive human male in the Epic of Gilgamesh.

Like Adam, Enkidu begins his story living naked and wild, in the natural environment, in balance with his animal neighbours. It is an advanced non-human female entity Shamhat who introduces the primitive male to more developed foods, to alcohol, to sexuality, to clothing and to city living. These are the roots of the Mowgli and Tarzan stories.

An oral tradition of the Zulu people tells of an advanced non-human female entity called Mbab Mwana Waresa. She introduces the distant ancestors of the Zulu people to the cultivation of crops and to beer. In Nigerian mythology it is, once again, the female who initiates a life of farming and self-sufficiency, raising the human condition from one of dependence on the advanced beings who hitherto have fed and nurtured them. For these primordial Zulu figures farming was their escape from being dependent on and controlled by non-human overlords. Through farming they become an independent, fertile civilization and begin to take over the planet as its new alpha species.

Genesis 3 – A New Kind of Human

This famous chapter in Genesis speaks of an upgrade to the human condition. As with the Sumerian and Nigerian stories of human beginnings, this chapter begins with an original cadre of humans who are not yet a fully-fledged, fertile species. Rather they have been directly engineered. The sexualization of the humans is accompanied by a cognitive upgrade in which the humans become self-aware, more intelligent and self-conscious about their nakedness, or maybe lack of body hair.

The genetic engineers in the story, whom the redactor has labelled Yahweh and the Serpent, are in fact stand-ins for Enlil and Enki in the Sumerian source narrative. In conversation, Yahweh and the Serpent summarize this upgrade in terms of the humans becoming *"more like one of us,"* by which they mean more like the advanced beings who are adapting them. Both the Biblical and Sumerian source narratives describe the fusion of proto-human and Elohim/Anunna DNA as the final step in engineering a species intelligent enough to be useful to their overseers without being so smart as to pose a threat.

Genesis 6 – The Flood

This is the Bible's version of the Mesopotamian deluge stories of Atrahasis, Utnapishtim and Ziusudra. It is a flood narrative which repeats all around the world, commemorating a bottleneck in the growth of homo sapiens, a global moment of repopulation post-cataclysm.

Genesis 11 – A Global Reset.

Known as the *Tower of Babel* story, this mysterious episode unfolds on the Shinar plane in the country we now call Iraq. When human beings arrive in this new territory, they do something they have not done before. What makes its first literary appearance on

121

the Shinar Plain is an advance in human civil engineering. It is advance so worthy of our attention that it is shared ubiquitously, having been memorialized in a story which I had always assumed was nothing more than a children's tale. However like so many *"fairy tales,"* which are often rooted in ancient traditions of male and female initiation, the message of this particular story now flies under the radar as a cute story for kids. I am talking about *The Three Little Pigs.*

James Orchard Halliwell, the British author and collator of *"nursery rhymes,"* put this story into print for the first time in the1880s, though the true origins of this story are shrouded in mystery, having existed for unknown ages as oral tradition. To refresh your memory, and in case you don't know it, the three little pigs live in a dangerous world of prey and predator and, unfortunately, they are not at the top of the food chain. At least, not when the story begins. For this very reason the little pigs decide to build structures for their safety. The first little pig builds a hut out of straw. Unfortunately this material fails to keep the wolf at bay who finds it easy to *"blow the house down."* The second little pig builds a structure out of sticks. When put to the test, it too fails to keep the dreaded predator out. The third little pig does something the other pigs haven't learnt to do. It lights a fire and then uses the fire to engineer a totally new material – brick.

With its fire-baked bricks this little pig then builds a solid structure and is able to defend itself and the first two pigs from the deadly predator. The fire also has other purposes. When the wolf attempts to attack this brick-built haven it is the fire that kills it, thereby providing the satisfying punchline in which the three pigs, secure in their brick home, sit down for a feast of cooked wolf. The prey has become the predator. The pigs are now the alpha species.

I would humbly suggest that in its original oral iteration, this story was probably not about pigs. Pigs do not light fires, engineer new materials or build structures. Nor do they sit at the top of the food chain. This is a story not about porcine progress but about human technology. It dramatizes innovations which were in reality major leaps forward in human progress, shelter-building, fire, weaponry, materials science, brick building and cooked food.

Fire changes a lot of things. Greek mythology marks this same leap forward in its account of Prometheus gifting humanity with fire, the seed of so many advances to come. The moment when we began our journey into materials science, and learned to menace and kill the animals that might otherwise have eaten us, as well as kill and cook our own prey, is a significant pivot in the human story. The advent of brick, en route to megalithic stonemasonry, changed history in a truly material way.

Together, the use of fire, and building with brick not only made human colonies defendable, but provided human culture with a way of leaving archaeological imprints which would long outlast the presence of the culture itself. This pivot, along with a note about its significance for posterity, occurs in Genesis 11 and contemporary archaeology confirms Iraq, Iran and Syria, otherwise known as ancient Mesopotamia, as the places where brick-built structures made their first appearance some time in the 8th millennium BCE. There is, however, another layer to the Babel story. Either it is the memory of a prior human civilization, or it is testament to another presence, a menacing, non-human culture, resident in the world of ancient Mesopotamia. But more of that a little later.

Genesis 12-18 – The Progenitors of Humanity

These chapters introduce the reader to Abraham and Sarah. It appears later in the current overarching narrative of the Bible, and at first Abraham and Sarah appear to be modern human beings living in a world of city-states and nomadic tribes but, as I mentioned before, there may be another more ancient storyline woven into the narrative and there are clues that this too is a human origins story. Abraham and Sarah, we are told, are unable to have children on account of Sarah's age. It is only after a mysterious intervention from non-human visitors that Sarah conceives, enabling the couple to become the progenitors *"of many nations."* The name Abraham, we are told, means *"father of many nations."* This title relates Abraham and Sarah not purely to the heritage of the Hebrew people, but to the origins of many peoples, nations and ethnicities.

However there is more to this couple, their names and their title, than meets the eye, because the Vedas of Hinduism introduce us to a mysterious, primordial couple whose names sound suspiciously similar: Brahma and Saraswati. The similarity goes beyond their respective names because Brahma and Saraswati are honoured as *"the progenitors of the many nations,"* meaning the ancestors of all modern humans.

Keep in mind that the earliest Hebrew script has no vowels in it, and then take another look at the similarity of the names Abraham and Brahma. In terms of consonants the names are identical: BRHM. In the case of *Sarah* and *Saraswati,* the word part *-swati* is a descriptor which means beautiful. *Saraswati* means *Beautiful Sara.* The names themselves are virtually identical: *Sarah* and *Sara.* (In English they are alternative spellings of the same name!)

I would suggest that this is no coincidence and that embedded within the Bible's Abraham and Sarah story is a yet more ancient narrative, a primordial story of human beginnings, in which this unique pair embody the genetic source not just of the Hebrew people but of the whole of modern humanity.

124

There is a telling post-script to the Genesis account of artificial fertilization which would support the idea of Abraham and Sarah as a human origins story. It comes in Genesis 26.

Genesis 25-28 – Human Beings become the Alpha Species

When we read these chapters in their current setting, the two characters of Jacob and Esau give the appearance of modern humans, two brothers, grandchildren of BRHM and Sarah, and offspring of the artificially conceived Isaac. However, when seen on its own terms and read as an independent narrative, another layer of meaning reveals itself. Cast in that light, we appear to be looking at two different kinds of human being, vying for pre-eminence in the environment. One is smooth-skinned and intelligent. The other is bigger, stronger, covered in hair as thick as a goat and rather less wiley.

This scenario echoes the story of human origins laid out in the Mayan Popol Vuh. It tells us that the Feathered Serpents' intervention in hominid development resulted in modern hairless humans and also *"the hairy ape-like creatures who live in the forest."* Here in Genesis, just as in the Popol Vuh, we have the hairless alpha and the hairy beta both resulting from external genetic interventions in the generations prior.

The Biblical account includes the fascinating addendum in which the smooth-skinned human now has to work out how to become the alpha and control the environment. Ultimately, he does this by learning to manipulate the hairy hominid with offers of food. This is how we tame wild animals. It is not the story of brothers.

For these reasons I would suggest that woven into the Hebrew story is a memory of the time when we lived among other potentially hostile kinds of hominid, when like the three little pigs we had to learn how to establish ourselves as the dominant species on the land.

Seven Catastrophes

Genesis 1 – The First Flood

As we saw before, planet Earth is not in good shape when we first encounter it. Shrouded in a sooty pall of darkness and flooded. This is the primordial state described by so many of the world's creation myths and is captured in the Hebrew phrase *tohu wa bohu*, which implies that some catastrophe has laid the planet to waste. In *Escaping from Eden* I suggest that this may be the planet's condition in the aftermath of an asteroid impact, the resultant sky fires, volcanism and flooding. The visual correlations expressed in widely different vocabulary and metaphor hint at the collective memory of survivors who, as a pre-technological species, try to describe the terraforming technology they witnessed in action. Hence we hear of the Iroquois *"turtle"* reclaiming land from the flood, and the Filipino *"hawk"* clearing the high land of floodwaters. In Genesis the helpers are the *elohim* and the technology is the *ruach*.

Genesis 3-4 – The Release of the First Modern Humans

These chapters tell the story of the releasing of genetically improved hominids (homo sapiens) into a world of non-upgraded hominids. Adam (a word which means *earthling* or *human*) and Eve (a word which means *the living*) are locked out of the enclosed zone in which the elohim have formed and nurtured them. The new humans will now have to gather their own food, survive without medication and as newly fertile beings, deal with the difficulties of childbirth.

On his expulsion from the region, Cain, the first killer, is terrified of the kinds of human beings he is going to encounter outside of the enclosed zone of Eden in which his parents were engineered. Released into the general environment, the upgraded humans must now compete with their neighbours and either sink or swim.

Genesis 6 – Another Flood or Two

This chapter of the Bible contains the fusion of two flood narratives. They appear to represent the same moment in which a rescue capsule safeguards five human bloodlines from the devastation of the flood in order to repopulate the planet post inundation. One narrative has been provided by a narrator who tells the story from a Yahwist perspective, neatly compartmentalizing the distribution of livestock according to Yahwist food laws. The other is the contribution of an author with no such concern about kosher regulations.

Genesis 9 – Planetary Revision

Genesis 9 hints at a terrestrial civilization prior to homo sapiens, a civilization before the dinosaurs, before the Cambrian explosion, in a forest of Giant Redwood Trees, a world of one ocean and one landmass, which we call *pangea*. It is hinted at by a verse at the beginning of Genesis 11 which suggests a single landmass with a single coastline. And it is puzzlingly referenced in Genesis 9 when the narrator recalls the separation of the lands. If there was a civilization on Earth at that time, we would have no geological record of it. By now all its artefacts would be ground to a powder. We would only know if some other intelligence had told us.

This possibility is put forward by Plato when he discusses people who live on islands in the sky, whose lifespans are far longer than ours and who have a better understanding of the true nature of outer space and the movement of objects through space. It is, he suggests, the movement of such objects which have caused periodic cataclysms on planet Earth, sufficient to take civilizations back to a virtual zero, erasing the memory of them and demanding a reboot of the whole ecosystem.

Plato doesn't define what these space objects might be, leaving open the question of whether such an object might be an asteroid, a missile or even a moon. With respect to the moon, the major scholarly consensus among astrophysicists is that our moon did not originate anywhere near this solar system but was captured at some point in our planet's history by Earth's gravity. Such a capture would have created catastrophic conditions on our planet's surface. Any civilization here at that time would have had major tidal and tectonic conditions to survive, if at all possible. Curiously the story of the moon's calamitous arrival is referenced in the Zulu story of Wowane and Mpanku who bring the moon to Earth from another part of space. Once in situ they hollow it out. In fact the story describes the moon as an egg emptied of its yolk – a curious reference to the moon's anomalous low density. How would ancient people know this?

Equally curious is that the moon's arrival is associated with cataclysms which extinguished the advanced civilization here at that time. More curious still is the Greek mythology of a previous, more advanced civilization in the deep past, whom they referenced as Arcadians. The second century Greek historian Pausanias wrote of a people group more ancient than the Athenians who described themselves as *Proselenes*. They saw themselves as the inheritors of the Arcadians who they said were here *"before the moon arrived."* So the hints in Genesis at a prehistoric calamity obliterating an earlier civilization is not without company in the annals of world mythology.

Genesis 11 – The Obliteration of an Advanced Civilization.

As we saw before, Genesis 11 speaks of a leap forward in civil engineering which we can correlate with the region's archeological record. For me something more intriguing still is that in recent times the word's defense forces have taken a great interest in the other layer of that chapter's story. It references something called a *bab-el*. The root meaning tells us this is a *"Power-Gate."* What is one of those? It could mean a *Powerful*

Gateway, it could mean a *Gateway for the Powerful Ones*, or a *Gateway to the Powerful Ones*. The story itself gives us an idea of which it might be. The Genesis text speaks of a structure being erected with the intent of *"reaching the heavens."* Once we recognize that *"the heavens"* is not some poetic reference to an imagined or idealized realm but is simply the ancients' way of speaking about space, that should give us pause for thought as to what this enigmatic structure really was. According to the Sumerian source narrative, the purpose of this enigmatic *Power Gate* was to dispatch hundreds of *"observers"* to their *"stations in the stars."* Do I need to say any more? You can probably do the math. When we picture the famed *"Tower of Babel"* we may imagine the familiar brick-built ziggurat structure, but the action of the Genesis 11 story is that of a stargate and we wanted to find the material remnants of it, we would send archaeological missions to the Shinar Plane in what is modern Iraq. Without going into detail, through my work among U.S. veterans I have come to understand that this is a story our friends in the world of military intelligence take extremely seriously.

As the story unfolds, this prior emergence of space-faring technology was not welcomed by certain elohim observing Earth from afar. Though the current redaction of Genesis 11 presents it all as a Yahweh or God-story, root meanings reveal a community of elohim who were determined to keep Earth isolated. In the same way that a modern empire-building nation might bomb another country's airstrip or airport for fear it could be used as a staging post for incoming foreign attacks, so the observing elohim decided to obliterate ancient Iraq's interstellar technology.

Then, by some unexplained modality, the same hostile elohim assault the neurological wellbeing of the previously spacefaring population to the point where they can no longer speak. Genesis 11 is not, as it is commonly preached, the story of the addition of new languages. It is the loss of capacity to speak intelligibly. Whether chemical, vibrational or radiological it was a neurological assault which consigned those ancient spacefarers, whoever they were, to a pre-stone age condition.

Genesis 18-19 – An Alien Attack

These chapters report other bombardments on the planet's surface. In chapter 18 Abraham encounters El Shaddai, whose name means *The Powerful One, the Destroyer*. A short chapter later El Shaddai has already obliterated two cities with technology which would have been mind-blowing to a pre-technological Abraham and his retinue. Small wonder that this story has made such a deep impact on our collective psyche.

Exodus 1-12 – A Series of Unfortunate Events

In the Biblical narrative of the exodus we have the story of a time and place visited by calamity: the failure of crops, plagues of locusts, the death of livestock, the infestation of the land with frogs, lice and insects, the poisoning of the water supply, an epidemic and the deaths of children. The tribes of Israel escaped all that calamity in Egypt in exchange for a life as nomads. Now, it is fair to mention there are serious questions surrounding the historicity of this narrative. The challenge has always been to find any kind of corroboration of this story in the chronicles of ancient Egypt, in the absence of which many scholars believe it to be an aetiological fiction. If later generations of Israelites were to ask, *"Why were our patriarchs nomads? Why were we so low down in the pecking order? Why did we not have land of our own?"* the story of the exodus would be produced as the answer.

We saw earlier that the Abraham and Sarah story carries a layer within it which can be placed in a far more distant moment in the human story, as a kind of Adam and Eve story. In a similar way, it is possible that the exodus story also carries a layer which has nothing to do with the timelines we know for ancient Egypt. After all, the text does not name any specific pharaohs in its telling of the story. Take a step back, and what we have is a story of calamity and escape. Painted with a broad brush, it is not unique in the annals of human history, as we will see a couple of chapters from now. Indeed, as you'll see later, my own family line carries the memory of a time of nomadic existence and an international migration, made in order to escape a plague-ravaged country. In that sense, when seen from a greater distance, as a story of calamity and escape, the exodus is a very relatable experience.

In the following chapter I want to take stock of these calamities and new beginnings and ask what they have to do with what certain members of the U.S. Congress are concerned about in the present. Do these Biblical and ancestral stories give the world today a reason to be concerned over what may be about to happen?

CHAPTER NINE

Threats and Compacts

Did you notice something alien about the fourteen traumatic pivots in the human story in the Genesis and Exodus accounts which we just surveyed in the previous chapter? Of those fourteen episodes, thirteen are associated with visitations from the heavens by advanced non-human beings, what today we would call *extraterrestrials*. Of the thirteen extraterrestrial visitations, only four are hostile, representing invasions or attacks. The remaining nine are benevolent. That's the ratio: roughly two thirds positive and one third negative.

Among the four hostile interventions, one stands out as especially relevant to today's conversations between Washington and The Pentagon. I am talking about the civilization-ending assault on ancient Iraq – an alien attack launched because the Shinar culture had begun its technological journey into interstellar travel. But wait a minute? Take a look at that. Isn't that us? Isn't that our situation in the twenty-first century?

If we are advancing on the research path towards interstellar capability, and if we are getting close to a breakthrough, could that be why we are exciting the interest of cosmic visitors today? Is that the reason for the acceleration of mass sightings around the world in the last twenty years? Could that be why the language of *"national security"* and *"existential threat"* is newly appearing in Senate briefings and Congressional House Hearings in Washington DC? Is this what our parliamentarians are really concerned about - a sabotage of our technology or an obliteration of our civilization, Babel-style? Do they think we are on the brink of another Genesis 11 takedown?

Of course it is possible that our members of Congress have not made these connections at all. They may simply be watching the Pentagon's video footage of the 2004 *Tic Tacs* and other UAP's, thinking that this is technology we ought to have if we want to feel secure and evenly matched in our corner of the galaxy.

There may be other reasons, besides, for this choice of language. It isn't hard to see how a notion of *"existential threat"* might be leveraged by stakeholders in military technology – the economic engine-room of the Military Industrial Complex which has driven U.S. foreign policy. Do we really want the same dynamic of profitable hostility driving planet Earth's relationship with interstellar neighbours with technology so much further advanced than our own? How wise would that be?

Unarguably, visitors with advanced technology such as has been tracked, filmed and recorded by U.S. Defense, would have to be understood as a *"potential"* threat. However, rather than blindly juggling hypotheticals, we need a grounded perspective on what the level of threat is in reality.

For this we must make a journey to Jerusalem to consider the revelations of a senior government spokesperson for Israel. For an entire generation, the person in question has enjoyed access to more privileged information with which to answer these questions than almost anyone else in the world. He is Professor Brigadier-General Haim Eshed.

Before his recent retirement, Haim Eshed was Israel's Chief of Space Security. It was a position he held for twenty-eight years. In that time, he headed up Israel's space program, among many other things oversighting the Ofek satellite program. His fundamental role was to know from moment to moment the nature and extent of any kind of strategic threat from space. Just before Christmas 2020 Professor Eshed announced that his understanding, informed by decades of privileged information, is that the USA and Israel are in contact with extraterrestrial visitors and have been for decades. These visitors are divided into various demographics who together comprise a *Galactic Federation.*

A galactic federation may sound like something straight out of the fictional canons of *Stargate, Star Wars* or *Star Trek,* but Haim Eshed has simply applied fresh vocabulary to a concept repeated by ancestral narratives spanning the globe.

Whether we are reading about the Feathered Serpents of China or the Yucatan peninsula, the gods of Greek legend, the airborne kings of the Vedas, the aesir of Norse mythology, the Apkallu of Babylonian and Sumerian texts, the Anunu of Hawaii, the Anunna of Sumerian cuneiform tablets, or the *El-Ba'adat* the *Council of Power* in the Hebrew scriptures, we are essentially looking at the same thing, a committee of competing non-human factions, bound by some kind of accord to manage *"Project Earth."*

In the El-Ba'adat of the twenty-first century, Ham suggests that unnamed other terrestrial governments besides Israel and the U.S.A. are in communication with our ET overseers at a covert level. He goes on to explain that compacts have been made with our visitors from the Milky Way, agreeing to a program of research experiments on and around planet Earth. Apparently, our earthly representatives, whoever they may be, have agreed to support our visitors' exploration of the fabric of the universe. Furthermore, Professor Eshed echoes Neil Armstrong's hints about hidden breakthroughs, and Ed Mitchell's assertions about embargoed technology, more of which a little later.

Most sensationally of all, the Brigadier-General asserts that American and Israeli special access projects already have hands on access to our visitors' technology, technology with interstellar capability and that *"we"* are already involved in collaborations

135

using that technology. Whether we are in a position to replicate said technology is a different question, and one we will return to a few pages from now. Critically, nowhere in Professor Eshed's statement to the press does he invoke the language of *international security*. Not once does he hint at an *existential threat*, noting that it was his job to know if there was one.

Moreover Haim Eshed goes on to assure us that on more than one occasion our visitors have intervened to prevent what could have been nuclear catastrophes. This is a reference to incidents during the Cold War in which nuclear warheads of both the US and Soviet arsenals were remotely activated and de-activated by an *"unknown"* intelligence. In this way, Haim Eshed's revelations indicate a non-human hand in our geopolitics, in the politics of war at the very least.

Reviewing Haim Eshed's words suggests a longstanding and a stability to these exopolitical arrangements and I note that he also identifies the same eighty-year timeframe referenced by the recent revelations of the Pentagon in Washington. If this compact with our visitors has been stable for eighty years, then it raises some important questions:

- Is this stability currently at risk?
- Is the recent acceleration of disclosure a hint that the watertightness of this secret compact is compromised?
- Might the factions involved be less agreed than in decades past?
- Has this stability lasted because extraterrestrial stakeholders already have all the power they need?
- Are our ET overseers comfortable with what they are already deriving from their stake holding in planet Earth? Or, to put it more bluntly, have we already been invaded?

The Brigadier-General, a man whose bona fides are beyond question, is not the first person to *"out"* this kind of compact or arrangement. In the 1600s a Presbyterian Minister, Robert Kirk, who served the parish of Aberfoyle, in the high country of the Trossachs in Scotland, wrote an extraordinary book called *The Secret Commonwealth*. In it he argued that no understanding of the world is complete until one recognizes that there is a non-human layer to the governance of human society. This non-human oligarchy, he argues, would appear to regard human beings in a way analogous to the way we would view livestock. Moreover, he argued, this mysterious presence is known to the human elites who appear to run the world.

Reverend Robert Kirk's ministry background was as a Bible-translator. Though well-regarded in his conservative Presbyterian circles some of his translation work was regarded as being *"too open minded,"* meaning that it lacked the familiar overlay of religious associations that the Presbyterian synod generally preferred. His open-minded translations of the ancient Greek and Hebrew texts of the Bible were only the beginning of his theological journey into the realm of contact phenomena. I can certainly relate.

A sympathetic pastor, Robert Kirk gave an open ear to the people to whom he ministered and among whom he had grown up. Gradually he became convinced that what he was hearing about a non-human presence, abduction, hybridization, and covert interference in patterns of government was something other than cute bedtime stories. They represented a coherent canon of contact phenomena which perturbed his parishioners. The locals spoke of *fauns*, *faeries* or *elves*. The word Kirk preferred was *Sith*, which he took from the Old English and Celtic words which indicated the world of the other beings. No need to ask what literature the scriptwriters of *Star Trek* and *Star Wars* were reading! Captain

Kirk is the iconic commander in *Star Trek* and the Sith are part of the pantheon of *Star Wars*.

Through my pastoral work with experiencers in the twenty-first century, I know that many who experience close encounters today also report unusual levels of precognition in the aftermath of these encounters. Kirk recorded exactly the same phenomena. He also described the anomalous visitors' anomalous ability to materialize and dematerialize, pinging in and out of visibility. He distinguished between the human-sized Sith masters and the child-sized and emotionless foot-soldiers. Two other details which are uncannily close to the distinctions of beings reported today.

Kirk recognized that his parishioners' reports were neither fables nor jokes. His people were genuinely concerned about an alien presence which many of them had witnessed, which many believed to be involved in abductions and many more of them understood to be influential at a covert political level. In this regard, Kirk noted that Sith appeared to be insensitive to human suffering and that their political influence could therefore be discerned whenever governments appeared to rule with no regard to the human cost of their actions. Given that, we might wonder if such a cold-blooded modus operandi should be looked for as the signature of covert, non-human influences in our politics today? Are these the clues that our elites are already under the sway of a *"secret commonwealth,"* a covert uber-government with no fellow feeling towards the great mass of humanity?

What does it look like when human society is influenced or governed over by *"Non-human Biologics,"* or an *El-Ba'adat* or a *Secret Commonwealth*?

In my book *The Scars of Eden* I argue that the Bible is laced with answers to those questions once we translate the texts according to the root meanings of the key words. In that light the chapters of

138

Genesis and the books which follow reveal the conflicts among the El-Ba'adat the following matters:

- How much access to information should regular humans be allowed? *(Genesis 3)*
- What access should regular humans enjoy to clean food and water supplies, medications and all the modalities of natural health? Should access be free, controlled or limited? *(Genesis 3, Deuteronomy 31)*
- How long-lived should we expect or allow humans to be? *(Genesis 6)*
- What is an acceptable number for Earth's human population? How many human beings should there be? Is a cull needed? *(Genesis 6)*
- Can we deceive an entire nation and trick it into invading or going to war with another group on entirely false intelligence? *(I Kings 22)*
- Should we take a population of human beings off-planet? *(Ezekiel 37, Revelation 21)*
- How much abduction and hybridization of human beings should be allowed? *(Genesis 6)*

Within that Council, whether we read the Biblical, Mesopotamian, Greek, Scottish, Norse or Mayan accounts of it, on the one side we see factions who acted for the freedom, health, longevity, education and technical progress of ordinary human beings, and on the other side are the factions, with less fellow-feeling towards humanity, who didn't want any of those things. I can't help feeling that the issues which divided the biblical El-Ba'adat sound eerily familiar. There is a strangely contemporary ring to them. So when I turn to today's global economics and geopolitics, I have to wonder, were our ancestors describing the past only? Somehow these issues are still live. In bequeathing their stories to us, were our ancestors really providing us with a lens by which to perceive

139

present dynamics and understand why things are the way they are today?

When questions like these are the issues of the day – the control of access to clean food and water supplies, access to and control of information, medication and other modalities of health, human population size and its possible reduction, questionable or proxy wars, and so on, our ancestors would say that these are *non-human* or *inhuman* agendas. To use the traditional Hawaiian language that I have learned from Kam, these are *ahumanu* agendas. If issues like these are on the table once again, could it be *ahumanu* influences are making their presence felt once more?

As to Professor Brigadier-General Eshed's references to *"research collaborations"* I wonder how they might relate to our ancestors' accounts of abduction and hybridization and their references to technology sufficient to take select humans off-planet. These ideas sound like pure Hollywood fantasy to the twenty-first century mind, but as I have learned through my travels, they are simply the stuff of indigenous knowledge the world over.

My personal journey of rediscovery began in Victoria Australia and has carried me to ancient Mesopotamia and to the Yucatan Peninsula, from there to the lands of my ancestors, Ghana, West Africa and Wales. Travels in Greece, Nigeria, Cameroon and Zambia all resounded the same notes. (You can share these travels with me in my book *The Scars of Eden*.) But more than any of my previous travels it has been my time with Kam that has best crystallized my sense of what our past invasions may have looked and felt like. His grounded perspective makes me less concerned by the idea of space invaders turning up again on the Whitehouse lawn and more concerned by what governance under ET influence might look and feel like. I certainly wouldn't want a repeat of what invasion looked like under the Mo-o, the Ahumanu, or the Anunu.

I would like to think that the relative stability of the arrangement we have had in place since the 1940's suggests that there is a healthy balance on that Council of Powers. In the twenty-first century we could certainly benefit from the support of pro-human demographics represented in antiquity by figures like Asherah, Mbab Mwana Waresa, Hun Hunahpu and others in that vein. But until further intelligence is released concerning the detail of the present *"collaborations,"* you and I can only guess and infer what the state of play might be.

Meanwhile, I am just having to get used to being regularly lambasted by commentators on *YouTube*, TV and radio shows stridently taking me to task for putting any of this speculation out there. Either I am a *"Suspected Illuminati,"* or a *"C.I.A. shill,"* or I am, as my correspondent Robert put it so elegantly, *"A Bolshevik and a psyop, luring people to the pit of hell."*

In moments like these I have to remind myself of the wise words of Oscar Wilde when he said that there's only one thing worse than being talked about, and that's not being talked about! To be honest, I really don't mind fielding the odd theological volley from friends like Robert, if it means that the bubble of taboo is burst and people are able to broach the question of ET contact both in the distant past and in the present. For this reason, just the other day, I was delighted to learn of a national gathering of Church of England clergy involved in paranormal ministry. The purpose of the gathering was to further develop expertise and professional standards in the spooky world of exorcisms, deliverance and entity removal, an element of my own work during my years as a church doctor. In the course of the event, the question of my work in paleocontact was raised and it emerged that the higher ups in this department of the Church of England

141

have lately been receiving a continual flow of letters from priests and parishioners asking for an official response to my *Eden* books.

"Does what this crazy Paul Wallis says have any truth in it? How do we respond to it? Is he right about the biblical narratives?"

"Is he right that the churches need to be ready for official disclosure of ET contact? Because as of right now we have nothing to respond with."

"The Pentagon has admitted that they already have possession of ET craft – and pilots. Where does that put our faith and our theology? With everything that's already in the news, why has no Church of England voice spoken up on this, other than Paul Wallis?!"

I have to admit that I am pleased to hear that these kinds of questions are being tabled and that my work in particular is being discussed at that level, although the prospect of powerful institutions weighing in, and potentially adding their lambasting into the mix does prompt a sharp intake of breath from time to time. Needless to say there is a certain cost to being a frontman for paleocontact. People's reactions can be awkward and sometimes less than friendly, even after decades of friendship and professional camaraderie. Of course I appreciate that it can be unsettling for devout people when sacred cows are put under the spotlight and unsettled people can be unpredictable. In fact, now I've mentioned it, I think this might be a good moment for a nice, calming beer, something which is really a must for any Australian author, and I am long overdue for a catch up with my theological friend, Brad. Brad is a friend of many years standing and he always peps me up with his *"ministry of encouragement."* It's an opportune moment for it and as there's a special deal on tonight at the Pacific Bay Bar and Grill, that's where we will head in the next chapter.

CHAPTER TEN

Plato and Paleocontact

The Pacific Bay Bar and Grill, Queensland - 2024

"I liked your books much better when you were writing about Christian spirituality instead of all this ET nonsense. I mean really, you're making yourself look as daft as a brush."

As I said, Brad is famous for his *"ministry of encouragement."* He's still got it, and of course I know what he is saying is true. What I write about is stretching and many people's initial reactions are going to be scorn and ridicule. Of course I know this, and Brad is allowed to point it out. He and I go way back. In fact, Brad has some theological training under his belt, too, so he understands a lot of where I am coming from. He knows that I have a strong theological case for my position relating to ancient ET contact in the Bible, and from time to time he gives me clues that we might even be on the brink of agreeing. Today, though, I can see he is eager to *"encourage"* me back onto the straight and narrow. He continues,

"You, my heretical friend, are going to lose all the lovely readers you've attracted over the last thirty years by putting out more Eden books. Why bring ETs into it and offside at least half of your hard-earned readership? Why upset people? Why not write another book like 'My Dinner with Anton.' I really liked that one. It sold well enough, didn't it? And it didn't offend anyone!"

Brad pauses for a few gulps of Heferweiser. What he says certainly has some truth to it. I wrote *'My Dinner with Anton'* when I was deep in my eastern orthodox period. At the time I was working as a priest in Kings Cross, which in those days was a rather rough part of London. I was in a Roman-rite parish within the Church of England and the parish uniform of black cloak and black cassock gave me a look like that of Neo in *The Matrix*. The addition of my very full beard gave me the appearance of a Greek or Russian orthodox priest, so much so that my older Greek neighbours would politely nod or bow whenever we would pass each other in the street. And I would always politely bow back. That's how eastern orthodox I looked, and my theology at that time would have been close to that tradition. *'My Dinner with Anton'* reflects that period in my life.

The book was about a Russian mystic hermit, a real person who lived from 1759 to 1836, who through a protocol of controlled conscious breathing had been able to activate prodigious cognitive powers – what Kam refers to as an *open manawa*. Seraphim's story, carefully recorded in print by eyewitnesses, shows in flesh and blood terms what an open manawa looks like. For instance Seraphim displayed a well-honed facility in what Kam calls *Maka hi'e hi'e*, what you and I might call *remote viewing*. Seraphim's facility extended to future sight or precognition, telepathic and empathic connection, and intuitive knowing, or *Na'au* to use Kam's word for it.

I have lived with Seraphim in my head for four decades and have done my best to nurture these cognitive abilities in my own life. This is something Brad has seen for himself and knows to be real. Brad is in fact a deeply spiritual person but, being an Aussie bloke, likes to pretend he isn't.

"So how's your manawa coming along, Paul?"

When Brad teases me I always like to give him a sincere answer, just to shame him.

"You may scoff, young Brad, but I know you've experienced the same metaphysical things I have. Any cat-owner or dog owner will tell you that their pets are conscious of things that you and I can't see and that they have a sixth sense which is more acute than ours. I think it's not so different with young children. They see and hear things that we adults either don't see or hear, or we don't allow ourselves to see. We filter the anomalies out."

"For instance, back when I was working as a Church Doctor, from time to time when I was invited into a troubled community, I would find that the children would tell me straight what everyone had seen and experienced in that community but which the adults might not even have mentioned, because they were struggling to process it and explain it to themselves."

"I think children don't feel the same obligation to be able to explain everything that's happening around them. So, often they will simply report what they have seen without any filter. They're the ones who will say, 'Nobody likes to go into that room,' or, 'All the pastors here go crazy,' or "All the moms here are sad,' or 'All the kids here get sick.' They can see it. They can describe it. Adults can struggle with that."

"I think that as adults our filters are strong. In my coaching, for instance, I have often come across a scenario where a group of let's say four people experienced a close encounter together back when they were fifteen. Twenty years later, two will remember it and every so often they will talk to each other and recount the story, almost to reassure each other that they really have remembered correctly. The other two, however, not only won't talk about it, but will get really angry if ever the first two bring the subject up."

"It's not that they don't remember. They don't want to think about it because either the implications of what they saw are too *disturbing or because they still don't know how to process it. So in taboo areas it isn't that the adults just don't want to talk about it. They haven't even allowed themselves to acknowledge what it is that they have seen."*

Brad presses his lips and furrows his brow, a sign of reluctant agreement. Clearly, I am making headway.

"When children tell us that they have encountered something inexplicable, positive, negative or indifferent, rather than brushing them off with 'It was just a dream honey,' or 'It was probably a trick of the light,' or 'You probably remembered that wrong,' ask them for some details of what they saw or heard. If they're puzzling over it themselves, perhaps ask them to draw what they remember. In this way they can learn to pay attention to the kind of sensitivity and extra-sensory awareness that they naturally enjoy as children, and which a lot of adults tend to shut down. That is one very simple way you and I can help the next generation to nurture an open manawa."

"And Brad I reckon if you sat with any friendship group or family circle and asked people to share their experiences of things they have experienced that they can't explain, your estimation of whether or not we've got company would change pronto."

"You know how I used to joke about my rare moments of heightened awareness? I used to say they were due to my 'priestly powers?' Well since I have been on this research path I have learned about shamanic and mystical traditions around the world and I have discovered that indigenous cultures everywhere have curated protocols intended to heighten their people's higher faculties and nurture the natural acuity that we have as children. In indigenous cultures this is mainstream."

Brad is still quietly nodding and after a thoughtful sip of Heferweiser he adds his imprimatur to what I have just said.

"Well, Paul, as Hamlet once said, 'There are more things in heaven and earth, Horatio, than are dreamt of in your philosophy,' which I think means that we should believe what we see even if our worldview isn't ready for it."

I like that.

"Bingo! Exactly!"

"I just think, Paul, that most people would expect something a bit more Christian-sounding from a venerable archdeacon such as yourself, something that sounds more like T.D.Jakes, the Archbishop of Canterbury or Father Ted, and a little less like Plato!"

To be honest, I don't think I should feel any need to apologize for sounding like Plato. Indeed I would take that accusation as a compliment. And, as I will show you in the next few pages, there is a lot more Plato in original Christianity than most people probably realize, and a whole lot more paleocontact in Plato then ever I was taught in school.

If you're not familiar, the ancient Greek thinker Plato is uncontroversially one of the most significant figures in the history of human thought. British mathematician and philosopher Alfred Whitehead once described the whole of the Western tradition of philosophical thought as *"No more than a series of footnotes to Plato,"* and nobody felt the need to challenge him on that. So when I quote Plato to my friend Brad, I feel like I am on a strong wicket. Brad might not take me completely seriously on the topic of paleocontact, but he surely can't brush Plato off as easily, can he?

"Brad, I know you think it's iffy when I get into extraterrestrials but what about Plato's take on this? Do you reckon he was 'daft as a brush? Because if you read his dialogues Phaedo and Timaeus and Critias, you will find that Plato talks about alien craft and their occupants. Fair play, he doesn't use the word extraterrestrials. He describes them as people who reside on islands in the sky. He makes clear that he regards them as non-hostile and understands them to be like us, only superior. He calls them people but notes that they are more intelligent and much longer lived than we are, and they have a more developed understanding of the sun, moon and stars and of deep space."

"Plato acknowledges the old idea of a flat Earth covered by a dome or firmament, but he says that beyond what we think of as the firmament is the true reality of what we would call deep space, and as for these advanced people on the islands in the sky, deep space is their area of expertise. More than that he says that we would get to see space the way they do if only we had the technology to fly and travel fast enough – which we now know is absolutely right."

"Plato claims to be writing Socrates' dialogues, so I have to say 'Plato slash Socrates says' that he has it on first person report that the Earth is actually a sphere, suspended in space, and that it resembles a leather ball, similar to a modern soccer ball, but coloured like a marble, swirled with white, blue and gold. Now, in our generation, we had no idea our planet looked like that until 1968 when William Anders took his now famous picture of Earth from Apollo 8 as he orbited the moon. For the first time you and I could see the Earth from enough of a distance that we could see it as it is, a globe suspended in space. And it looked nothing like we expected. We had all imagined a blue and brown map wrapped around a globe. The swirl of white, gold and blue, like the brushwork of an artist, is just as Plato described. So, then, who

told him that? And how do they connect with these people on islands in the sky, who know all about deep space?"

If you have read *"The Scars of Eden,"* you will already know all this about Plato's prodigious knowledge, but don't jump ahead. Stay with me because I am going to go more deeply into Plato's paleocontact than we have before, and by the end of this chapter you will realize that Plato, Socrates, the apostolic writers of the New Testament, and the early Church Fathers were well and truly across the idea, not only of ET contact in the deep past but specifically of a time when aliens invaded.

"So Brad, I am convinced that Plato-slash-Socrates was a contactee and I'll tell you why. He is quite convinced that in altered states of consciousness, like after a psycho-effective tea, Kykeon, in his case, or in some other near-death ceremony, or at death, people have close encounter experiences. By this I mean contact experiences and communication with other kinds of being – spiritual or interdimensional beings. In fact, Plato is a little bit vague as to whether these loving beings may be humanity's ancestors, higher aspects of ourselves, or something unimaginably different."

"But there is so much interiority in how he talks about this mysterious kind of encounter that I reckon both Socrates and Plato had this experience – whether in the Eleusinian sect or one of the other groups like the Cult of Orpheus or the Cult of Hecate. Certainly they were initiates of the Eleusinian sect, at least, which curated psychedelic protocols and near-death states. I am sticking my neck out, but that's what I would argue, and I know there are serious academic scholars of Plato who would agree with me on that."

"On another front, you've heard me talk about ancient visitors adapting human life and upgrading our cognitive abilities. I found

149

that story in the Bible in Genesis 3, in the Sumerian source narratives, and the Mayan story of the Progenitors. Plato has that too when he talks about the children of the gods. Within the Greek tradition, Prometheus would be an example of a child of the gods intervening in human development to upgrade our level of intelligence and technology. Plato was totally on board with that concept."

Don't be misled by the religious-sounding word, *'gods.'* It's really a very misleading word to use. When you and I hear the word *'gods'* the associations that click into place for us are *fictional, religious, eternal, non-material, imaginary, transcendent beings.* That's not what Plato means, and that's why I think we should be very careful of that word.

Plato is really helpful on this point because he explains very carefully what he means by that word. He says there are two kinds of use of the word *god.* Firstly it can be applied to interpretations or personifications of the planets, like Saturn, obviously. The second use, and this is the one he is interested in, indicates certain living beings, creatures more powerful than us, who live in or come from the heavens, from outer space. Today, we have a different word for that kind of living being, don't we?

"All around the world, ancestral narratives of human origins begin with the appearance on Earth of advanced beings who arrive here from outer space. The Zulu people carry the story of Unkulunkulu arriving from space. The Efik people of Nigeria speak about Abassi and Atai who live on an island in the sky. (There's that metaphor again!) They then manage human development and genetically modify our ancestors for higher intelligence."

"The Dogon people from Mali, West Africa speak about the arrival of the Nommos from Sirius C who gave them all their

ancient wisdom. The Vedas recall the ancient 'kings' arriving in their temples and cities in the sky. The Bible talks about the Sky Armies, the Seba Hassamayim."

"Plato talks about the demiourgos, this other kind of being, not a human being, who arrives in an already existing world and adapts it, basically creating the ecosystem that we're all familiar with."

"If you look at the first chapter of Genesis, or its Sumerian source the Enuma Elish, or the Mayan story of the Progenitores, the Tagalog narrative of the Philippines or the story of Osanobua and his sons, told by the Edo and Yoruba people, all those narrative traditions describe advanced beings arriving from space, and landing here in an already existing world, specifically at a time when the Earth had been flooded. That's an amazing coincidence – if you can call it that!"

"What I want to tell you today is that Plato's version of that interstellar arrival frames it as a conquest. He tells it just like it's told in Deuteronomy 32 and Psalm 82 where, El Elyon carves up the Middle East and parcels it out to Yahweh and the other the elohim for them each to have a territory to manage. Plato doesn't describe it negatively, but this act of division for management among the advanced beings clearly indicates that superior beings have just turned up and are taking over to divide and rule. It's an invasion."

Brad is used to my long speeches and he has listened manfully to my latest peroration on my Athenian friend.

"So, Brad, either you have to tell me I have totally misinterpreted Plato, or if you give credence to Plato, and I know you do, perhaps you ought to start showing me a bit more respect."

Brad laughs in the face of my challenge, spraying me with beer-foam in the process. He knows I am teasing him. The fact that our views don't entirely align has never dimmed this particular friendship, and for that I am very grateful. Agreeing to disagree is a skill worth treasuring. Having collected himself and mopped himself down, Brad places a thoughtful hand on my shoulder and gently explains why I am in the wrong.

"The difference, my heretical friend, is that Plato wasn't a Christian let alone a senior churchman. He was four hundred years too soon. So he can say whatever he likes. But what does any of that have to do with Christianity? You were a venerable archdeacon only a few short years ago. But today you sound like you've converted to Plato!"

Now I am leaning forward in my seat. I am ready for this question.

"You think I sound like a Platonist? How can that possibly a surprise? Surely you would have to say the same thing to the first Christians. Certainly it was true of the Apostolic writers."

Now he's really got me fired up.

"What do you think? Of course I am a Platonist, and so is every Christian in the world today! If you didn't have Plato, whole chunks of Christianity would totally evaporate. For instance, think about the Gospel of John's explanation of Jesus as the word, his pre-existence, his incarnation, and his return to the source of the cosmos, the Father, what the first Christians called 'the great repose,' that's all Plato!"

"Lose Plato and you've robbed the Apostle Paul of some of his most memorable punchlines. There are about a dozen famous quotes attributed to the Apostle Paul where all he has done is take a Plato quote and a change a word to give it a new spin."

"The idea that at the heart of our soul journey on Earth is to learn to shake off heavier emotions like rage, resentment, unforgiveness,

trauma, selfish ambition, anger, fear and attain to eternal wisdom, that's pure Plato."

"Or how's this: Paul's idea of an immortal soul, shaking off the mortal body, or the idea of transfiguring contact with ancestors and other spirits, I am thinking I Corinthians 15, Matthew 17, Mark 9, Luke 9, Hebrews 12, II Peter 1 and I John 4 in the New Testament. That's Plato."

"Think about the idea of a life-review after death, the hint of reincarnation, the notion of God as pure intelligence or pure consciousness, and that being the original source of the material cosmos. That's Plato."

"Take those philosophies out of the gospel of Thomas, or John or Matthew, or Paul, or Peter, and what have you got left?"

"The early Christians and the apostolic writers who wrote for Jesus, where do you think they got these philosophies from? Not the Old Testament, the Hebrew scriptures. Those ideas aren't there. They got these notions from Plato. Justin Martyr, Origen, Irenaeus and Clement of Alexandria, all affirmed Plato and the Stoics as the best and broadest preparation for what Jesus was on about."

"Just let that sink in a moment, because they had all read Phaedo, and Timaeus and Critias. They had all the information I have shared with you from Plato about close encounters, ET interventions, ancient cognitive upgrades, non-humans terraforming our environment and even alien invasions, and they didn't have any problem with it. So don't think for a minute that by calling me a Platonist you are offending my Christian credentials. You're setting me in the context of some of the greatest church fathers."

I am not sure that I have convinced Brad of the merits of paleocontact theory, and we conclude our conversation with Brad still convinced that I am selling out Christian orthodoxy, though at least he can be reassured that what I am putting forward, two thousand years ago, would have put me in good company.

One of the many things that inspire me about Plato is what he chose to do with his knowledge of cosmic interventions in the human story. His experiences of altered states convinced him on a visceral level. The fact that he was convinced of it intellectually shows both in his lightest of references and in his most closely reasoned arguments. Where then did this conviction lead him?

Plato could have chosen to throw himself into political life in order to change the official narrative and hopefully sow more enlightened thought into the ways of government and public life. Instead, like his mentor before him, Plato chose the way of education, devoting himself to a select group of students while at the same time investing his creative energies into writing literature to elevate any who would read his works from that day to this. Plato's hope was that a more self-conscious and intelligent populace would lead ultimately to better politics and a better way of life for human society.

Plato's legacy in literature has invited readers in every generation since to honour their curiosity, observe objective data, and then apply logic and reason to those observations, even and especially when we don't at first understand what we are seeing. Plato called this *"Sophia"* or wisdom. It is the foundation of what we call science, philosophy, and personal development. For him all these pursuits were of a piece. They were all aspects of human progress and ascension. In addition, Plato sows into his work a subtle advert for altered states of consciousness as a pathway for more intense and intuitive knowing, what Kam calls *na'au* and what the Hermeticists called *nous.*

Plato was a pioneer in many ways and a great synthesizer of world thought and, as I have learned in my travels through the wisdom of the ages, he was far from the last public figure to see the clear connection between paleocontact, contact in the present and human potential. This connection is something you and I will touch on a couple of chapters from now as we get into conversation with a certain source close to an American president whose name has long been associated with rumours of ET contact. Join me in the next chapter where we will rendezvous with the president in question on a pleasant golfing holiday at a luxury resort in Palm Springs, California.

CHAPTER ELEVEN
Alliances and Sacred Cows

February 24th 1954

It is the middle of the night in California, and a little more than a year into his term in office, President Dwight Eisenhower is on vacation at Smoking Tree Ranch, a resort in Palm Springs. But tonight the president is not sleeping. Under cover of darkness, a security dispatch has ferried him from his holiday accommodation to a nearby Airforce base. A fleet of dark sedans travelling together at two in the morning is hard to miss in 1954 and it isn't long before an explanation has been inferred from local eyewitness testimony. Accordingly the Associated Press releases a report announcing that the president had suffered a heart attack and died. Within minutes, however, the story is retracted and corrected with an official statement announcing that the reason behind the president's clandestine visit to the Airforce base was a matter of some urgent dentistry. Apparently, a crown had come loose and needed re-attaching. In the middle of the night.

However, this is not where the mystery ends because from February 1954 until the present day, sources close to president Eisenhower have leaked another explanation for this night-time foray. Far from an urgent dental need, the story has been told that the president's midnight meeting was with representatives of extraterrestrial factions and moreover that he was negotiating the terms of a treaty.

When I first heard this piece of urban legend I was skeptical. Purely on the basis of logic, it seemed to me an unlikely scenario. After all, what power could we as a species possibly have at a negotiating table with beings advanced enough to get to Earth in the first place? The vast scale of interstellar space would require such visitors to have mastered subspace or wormhole technology. What possible leverage could we have sitting next to visitors with that level of technology? The mismatch of power would be laughable. Surely, any agreement made at that table would be like a mouse agreeing with a lion that the mouse is going to get eaten. It would be a surrender by any other name. However, if we accept the account former Prime Minister Dimitri Medvedev, and former Israeli Chief of Space Security, Professor Brigadier-General Haim Eshed, then we are looking at something other than a binary conversation. Rather we are looking at a table occupied by multiple delegations. Haim Eshed's phrase *"Galactic Federation"* probably gives us a fair idea of what Eisenhower found himself presented with seventy years ago. Given this multiplicity of involved parties, it would be reasonable to infer that some matrix of agreements concerning the management of project Earth must already have been in place by the time Eisenhower was introduced to the world of exopolitics.

Given this, any exopolitical decision would have to be far more complex matter than a simple surrender. Players at the American end of the equation would have to calculate how best to position American and global interests relative to the various parties already in conversation with each other.

Logically, each faction would present with somewhat differing agendas – a notion amply reflected in the annals of the Bible's El-Ba'adat, and the Sky Councils of ancestral narratives. According to our ancestors, the motivations of our ancient visitors were diverse. Some elements came to our planet for the sake of humanity's survival and progress. Some came for the mining of minerals, others for human resources, and still others for human adaptation. If the past informs the present in any way, then any American calculations concerning who might be our best allies would have been complex indeed. Feeling out which demographics had humanity's interests closest to heart would only be one part of the equation. Other factors would have to be weighed and military intelligence would have been on a steep learning curve in assessing which factions were trustworthy, which might have hidden agendas, which were best equipped to understand our interests and which were best able to respond to hostility from others. Without a doubt, some subtle exopolitics would have been called for.

However, before we assume that these considerations were suddenly in the hands of President Eisenhower, and all to be settled within a single dental appointment, I want to suggest that Eisenhower was probably not the first American president to sit at a table with non-human visitors nor was he the first to have been briefed with precise information about their presence. There are many reasons why I think it likely that this *"dental appointment"* was part of a process of reading the new president into a web of decisions which had already been made.

Firstly, it was in the previous presidential administration that UFO incidents first spiked in the immediate aftermath of the USA's detonations of atomic bombs over Trinity, New Mexico, followed by the obliteration of Nagasaki and Hiroshima, Japan. These traumas were followed by a flurry of mass sightings of UFOs, which peaked in 1952. From the moment of the test detonation at Trinity on July 16th 1945, until the incident at Roswell on July 8th 1947, UFO incidents in American airspace were responded to publicly, by press, politicians and military alike. The gist of the official response was always, *"We will find out what these flying saucers are and when we find out we will tell you."* This was the official line right up until the day after Roswell. Then, in an instant the official policy from the top down changed from open pursuit of information to official silence, debunking, gag orders

159

and even death threats, which were issued by senior military officers to any local witnesses who might otherwise have considered opening their mouths concerning what they had seen. Why the sudden change? Logic would suggest that the pursuit of information had succeeded but that the information, once obtained, was decided to be not for public consumption.

This policy of silence, denial and debunking remained forcibly in place until 2019, at which point, pressured by the 2017 Mellon leak of the Tic-Tac footage, the Pentagon finally acknowledged that throughout the seventy-two years since 1947 it had continually maintained units devoted to analyzing materials from UFO retrievals.

Back in the 1940s the official shift from openness to censure was concretized by the National Security Act, which lay the legislative foundation for the CIA and all its mechanisms for managing information and maintaining levels of secrecy, modalities which ultimately would exclude even future presidents from knowledge of The Program on a need-to-know basis. Building on the protocols established surrounding the Manhattan Project, the new National Security Act developed a legal basis for the kind of security clearances which today keep secrets about exopolitics far and away from mere senators and members of congress, hence all the furore around the Grusch complaint of 2023.

Conspicuously, the National Security Act became law within two and a half weeks of the Roswell incident, and was effected with such urgency that it was signed into law off the planet's surface, mid-air, aboard the Airforce One of the day, an aircraft aptly named *"Sacred Cow."* The president pushing the sacred pen aboard the VC54C was Harry Truman. For all these reasons I would suggest that the modern era of exopolitical treaties was already underway prior to Dwight Eisenhower's administration, and that the real penwork which established the alliances and compacts which have lasted until now was done by military intelligence units serving the Truman administration, under the veils of secrecy codified and legislated by that administration.

Nothing in my education prepared me to regard the stories surrounding presidents Truman and Eisenhower as anything other than pure urban legend, something worthy of an episode on *the X-Files*, but nothing more than that. Even after my pursuit of root meanings revealed non-human entities like *Dagon, Asherah, Chemosh, Milkom, Baal, Beelzebul, Akhekh of Egypt, El of Ekron, Yahweh, El Shaddai, El Elyon,* and the other *benei elohim* in the pages of the Bible, a deep current of skepticism still ran through my mind with regard to close encounters in recent history. Somehow I found it easier to conceive of contact in humanity's deep past than to give credence to accounts of contact in the modern era.

However, in the years since publishing *Escaping from Eden*, I have followed the logic of paleocontact from the realm of global folklore well and truly into the present day. In that time my profile as an author and as the host of *5thKind.tv* has privileged me with a steady flow of information, person-to-person, which has progressively erased any sense of separation between the credibility of ET contact in the deep past and the experience of similar contact in the present. This flow of information has included the testimony of individuals within the Apollo program, senior military personnel and a few pages from now, sources close to President Eisenhower. Meanwhile I want to return for a moment to December 3rd 2020, a day which speaks directly to the credibility of the 1954 story relating President Eisenhower to the world of exopolitics, and to my proposition regarding the administration of Harry Truman. This was the day on which our friend Brigadier-General Haim Eshed sat down with Raanen Shaked and Gabriel Beharlia of the Israeli national newspaper,

Yediot Aharanot, and dramatically added flesh onto the bones of this seventy-year old story.

Yes, says the Brigadier-General, a compact has indeed been established with ET visitors. Yes, we have agreed terms which allow certain visiting factions to pursue their own *"research"* on and around planet Earth, and yes, as a result of this compact we already have covert access to interstellar technology.

This assertion supports the claims of the forty witnesses interviewed by David Grusch on behalf of the Geospatial Intelligence Agency, AARO and the U.S. Congress, witnesses who have now given sworn testimony to the Inspector General of the Intelligence Community. Under oath, they have brought first-hand report not only of having handled metamaterials derived from UFO retrievals, but of having examined entire craft.

Haim Eshed's statement went still further, claiming that covert contact with our ET visitors has extended to active *"collaborations,"* not only on Earth, but also on our planetary neighbour, Mars. These agreements, according to the Brigadier-General, include representatives of the governments of the U.S.A. and Israel, and extend back *"more than seventy years."*

More than seventy years? To be clear, that means prior to 1950. That's the administration of President Harry Truman. The moment we give credence to Brigadier-General Haim Eshed's statements to the press is the moment we have to accept that the urban legend surrounding President Eisenhower can no longer be dismissed, and that the great likelihood is that his involvement was secondary to decisions already taken under the prior administration of President Truman. If the scenario described by Haim Eshed is true, then it raises some important questions, not only about the past, but about the present:

162

- Who represents humanity at the table of galactic powers?
- In whose interests are decisions being made?
- What are the various intentions of our visitors?
- How active a role do our visitors play in our geopolitics?
- Our ancestors spoke of human abduction by non-human neighbours? Is that a present reality? Is it an element of the *"research"* that Haim Eshed references?
- What does Haim Eshed's language of *collaboration* really mean? Is it a polite word for *oversight* or *annexation?*

Inevitably, any secrecy surrounding deals which govern all our lives will give rise to suspicion. This is true at a purely human level. In Australia for instance we, the people, are not allowed to know the terms of our country's trade deals with certain other world powers. Why would that be? Logically, it must be because the terms of those agreements would not be popular or might even scandalize the Australian people. Why else would they be kept secret? What purpose then has been served by the cover of secrecy over exopolitical treaties – if they exist in the way indicated by Brigadier-General Eshed? Might the general population be scandalized by what has been agreed to at a planetary level?

In his mature years, Apollo Astronaut, Dr. Ed Mitchell, the sixth man to walk on the moon, campaigned tirelessly for the U.S. government to lift the veil of secrecy over the realities of current contact. On the basis of whatever privileged knowledge he possessed, but was not authorized to speak, Ed Mitchell's belief was that such transparency would only help the human side of the equation, and that it would be better for humanity at large to know who represents us on that council and what decisions are being made.

A significant aspect in Ed Mitchell's long campaign for transparency was his understanding that our current alignments on that council have made potentially world-changing technology available to us, specifically zero-point energy and gravity manipulation, technologies which would make today's energy markets obsolete. We could liken these technologies to Nikola Tesla's approach to drawing upon our planet's own electro-magnetic field for free power, with gravitic manipulation and zero-point tapping the energy of fields which are endemic to the cosmos.

In the decades following the Apollo 11 mission, Ed Mitchell's fellow Apollo veteran, Neil Armstrong, was a private and reserved person, not one given to speeches or extravagant claims. So it was an exceptional moment in 1994, at the twenty-fifth anniversary commemoration of his 1969 mission to the moon, in the context of a closely scripted and enigmatic public statement, when Neil Armstrong used a form of words which captured people's attention around the world. He said,

"There are breakthroughs available..."

Not *"There may be if this that or the other."* Not *"There will be at some point in the future."* No. Breakthroughs *are available.*

Available to whom? Neil Armstrong's answer:

"To those willing to peel back one of truth's protective layers."

What can this sore thumb of a phrase possibly mean if it doesn't refer to getting breakthroughs declassified? I don't want to put words in Neil Armstrong's mouth but I would suggest that his understanding of what those *"breakthroughs are"* and how and why their availability has been veiled by *"protective layers"* was probably not very different to that of his comrade, Ed Mitchell. Certainly if you watch video recordings of Dr. Mitchell in action on this topic, you will quickly ascertain that there was no doubt in his mind that zero-point technologies, with everything they have

164

to offer humanity, *are* already *available* by dint of present contact and collaboration. And note, that is what he was allowed to say.

Ed Mitchell's passionate view regarding embargoed technology was something he shared with the British whistleblower, Gary McKinnon, for whose protection the United Kingdom changed its extradition relationship with the U.S.A. back in 2011. Both considered it a scandal that so much human misery has been created by the high cost (for some) and inaccessibility (for others) of fuel and electrical power. So much of our monetary cost of living and so many of our negative impacts on the environment derive from the industries of oil, fossil fuels, and the generation of electrical power, whether by fossil fuels, cobalt mining or nuclear fission. Their frustration, and mine, is that none of these costs are necessary, if *"breakthroughs are available"* through our covert collaborations.

In Australia today, households are struggling and some families are losing their homes because of inadequate supply and therefore extortionate costs of power. For many households, their power and fuel bills far exceed their rent or mortgage costs. Meanwhile, on the opposite side of the world, in Great Britain, according to its own Office of National Statistics, on a conservative estimate, the winter months of 2020-2021 saw more than 8,500 British citizens freeze to death in their homes, dying of hypothermia because they were simply unable to afford the gas or electricity needed to heat their homes.

To give you a gauge as to why, in Great Britain 2022, for an old age pensioner to meet the average heating bill they would have to part with 75% of their state pension, which for some elderly Britons is their only income. If it were revealed that the annual death toll in Great Britain due to unnaffordable power was in reality totally unnecessary, and that a decision had been taken by the powers to embargo available technologies for free power, wouldn't you want to know who had made that call and why? Wouldn't you be asking whose interests are served by keeping this kind of technology under wraps? Is it naive to suggest that the happiness, health and life and death of human beings might be more important than whatever those other interests might be?

It was for reasons like these that Ed Mitchell and Gary McKinnon made their respective calls for transparency regarding our covert collaborations. They wanted accountability. Similarly, it is the withholding of information which lately has so incensed members of U.S.Congress. Members of Congress don't like being considered too junior in the pecking order to be trusted with intelligence held by the military intelligence community. David Grusch's complaint is that the kind of secrecy now being exercised within U.S. military intelligence goes beyond any protocols supportable by existing legislation. By going through all the correct protocols with DOPSR and the Inspector General of the Intelligence Community, David Grusch has not only placed himself beyond reproach, he has legally brought this constitutional problem to light.

Naturally, Congress now wants to know what it isn't being told, and in whose interests it is not being told. And increasingly I am sensing that the public wants the same thing. If the U.S.A. and, by implication, the world is facing an existential threat, how can that not be everyone's business? Anyone with a democratic bone in their body should be asking questions like these. Furthermore, if we are in contact, who is representing humanity in any discussions?

"Well, I can tell you it's not the politicians. They are not involved in this."

I am on the line again to Richard Dolan, this time interviewing him for *5thkind.tv.*

"Politicians are not even involved in governing their own societies any longer. Who really believes that? Certainly the politicians have no control of this sensitive subject."

I have to agree. Richard goes on,

"We have a very opaque system of power in this world which masks itself as democratic. It's a fiction. So what is it? Is it an oligarcy?"

I am fascinated to have Richard on this political track so I press him further as to exactly who the oligarchs are.

"I think some have attended Bildeberg meetings and World Economic Forum Meetings. Those are the kinds of people we are talking about. Some may be too big to attend those meetings. It's difficult territory to get into because there is no public discussion of this – and for obvious reasons."

"It's like America in the 1940's. Nobody ever spoke about the Mafia – even though it was clearly there! The rule was 'Don't discuss these things because it would implicate too many powerful people!' We are in exactly the same place with global governance today."

I ask Richard how this relates to whatever technologies may be emerging from The Program. Who are the stakeholders and gatekeepers in that department?

"We get a hint of an answer to that in the Davis-Wilson notes. We're talking about big money, private corporate money, major financial players who get their hooks into national governments and control them in the areas where they want to determine policy. This has been going on for years. They also have their hooks into the issue of UFO technology."

If you haven't heard of the memo that Richard mentioned, the *"Wilson-Davis Note,"* don't worry. I will share a bit more detail concerning that amazing document in the next chapter. If you want an insight as to how longstanding the issue is of U.S. intelligence agencies not sharing information regarding *The Program* even among themselves, then the Wilson-Davis note is a document you are going to want to read. As our conversation progresses, Richard gives me his view as to when he thinks this current polity of secrecy and exclusion might finally lift.

"I think the real key to humanity as a whole joining any kind of group that's out there will be when we master spacetime manipulation."

Now myy ears really prick up, because this is exactly the same bar indicated by Brigadier-General Haim Eshed in his book *"The Universe Beyond the Horizon."* This was the publication he released at the time of his press-conference with *Yediot Aharanot* back in 2020. Haim Eshed's understanding is that our ET overseers have decided to hold back on disclosure until such a time as human beings have finally come to understand *"what space is and what spaceships are."* This makes perfect sense. They are waiting for us to master spacetime. When our Department of Defense witnesses report how visiting craft appear to *"ping"* into our airspace from out of nowhere, or instantly accelerate and decelerate without crushing the *"non-human biologics"* at the helm with G-forces, or how they skip across the sky rather than moving in a smooth, linear manner, it strongly suggests that mastery of spacetime is exactly what our visitors have been able to do. So how close are we to unlocking the power of spacetime manipulation? Logically this is the vital technology we need if we are to open up the final frontier.

Given the vast distances which make Newtonian travel between interstellar neighbours all but impossible, the ability to play with spacetime, to skip and ping, in the ways we have observed, is essential before we can even escape the confines of our little solar system. By Haim Eshed's account, our observers are watching for the announcement of this technological leap forward. They are waiting for us to become spacefarers, and in recent months Jacques Vallee, Eric W. Davis and Garry Nolan have been allowed to reveal to the public, especially when asked the right questions, that we are getting closer. If they are allowed to say this much publicly, we have to ask how much further forward we might be behind closed doors.

Meanwhile, David Grusch is facing reprisals for his public challenge to the Pentagon. That's why in the next chapter we will head from Palm Springs to Arlington, Virginia and into the corridors of military power.

CHAPTER TWELVE

Endgames and Fiefdoms

Arlington, Virginia – August - November 2023

"We will not back down. Those of us who have supported Mr. Grusch and who stand beside him and behind him understand that Mr. Grusch has an incredible amount of integrity. At each part of this process he has done exactly what should have been done procedurally and ethically. [His] career in the military and intelligence community...[is] unimpeachable."

We are in the company of retired Navy Chief Master at Arms, Sean Cahill. Sean served in the United States Navy from 1995 to 2015 and was part of the U.S.S. Nimitz airborne crew which engaged with the now famous Tic Tac craft of 2004. The reason he is defending David Grusch is that certain factions are giving him pushback after the filing of his complaint to the Inspector General of the Intelligence Community and the presentation of his sworn testimony to Congress. Medical documents passed to *The Intercept*, an American podcast and online magazine, have been published with the apparent intention of undermining the public's confidence in Mr. Grusch.

The leaked document is an extract from David Grusch's personal medical records. The highlighted section concerns David Grusch's use of drugs to self-medicate for P.T.S.D. The issue was so minor that the medical intervention to support Mr. Grusch in tackling his P.T.S.D. made no impact on his career whatsoever. David Grusch retained his top security clearances and continued his career in intelligence. Yet the author of the *Intercept* article considered this official affirmation of his sanity and reliability, post examination, as irrelevant. The idea seemed to be, *"If we can smear David Grusch as a junkie then we no longer have to take him seriously."*

Whoever the individual is who authorized the release of David Grusch's medical records, it is clear that somebody, somewhere is extremely unhappy that the compartmentalization of U.S. intelligence community is being outed as a matter for public discussion. It is an embarassment because even from a distance, it is easy to see that this issue creates a significant vulnerability in terms of security. Cultures of compartmentalization within the world of intelligence created significant threats to the safety of passengers travelling the world on commercial airlines during the 1980's, the era of air terrorism. This was the context of my father's work in international security for three decades. So I am somewhat familiar with the vulnerability that such a culture creates.

However, for those within the intelligence community, compartmentalization can look quite different up close. After all, the whole *raison d'etre* of an intelligence agency is to find and keep secrets. Given this basic fact, if units within agencies and agencies within the community habitually withhold intelligence even from each other, how likely is it that the units with the biggest and most closely guarded secrets of all are suddenly going to start unburdening themselves to the U.S. Congress? Surely that's not going to happen.

With that as a given, some of my correspondents have been asking me why Thomas Monheim, the Inspector General of the Intelligence Community, was willing for David Grusch's complaint to be heard so publicly and to escalate into a Congressional hearing. Could it be that the real intended outcome is for oversight of The Program, access to its information and control of its black budgets to pass from one authority to another, on the pretext of *"transparency,"* but all the while keeping its contents carefully under wraps? Is it really no more than a political play intended to transfer The Program from one fiefdom to the next?

Whatever the intended outcome, somebody is unhappy that the light of publicity has been allowed to shine on an internal squabble and they seem to hope that if David Grusch can be discredited then the whole problem can be made to go away. However this personalized move against David Grusch has fallen flat as Sean Cahill and other comrades of David Grusch are now stepping up to the plate to call foul. Sean Cahill continues:

"The Intercept showed no interest in investigating the claims made by a whistleblower but instead went after him in a personal fashion...I do not think we will be going backwards. This investigation will continue regardless of what skeptics or bad actors may try to do."

Like Sean Cahill, I am not unfamiliar with bad actors in the workplace willing to stoop low to protect their own interests and advance their personal agenda. And, like many pastors, I have experienced episodes of P.T.S.D. myself. Indeed for many years I lived with a degree of agoraphobia, the result of a workplace trauma. So I have an instinctual empathy with David on that account and naturally hold a dim view of the kind of person who seeks to leverage something like that in the workplace. Certainly, I know that David Grusch was ready for this kind of retribution and that it came as no surprise to him. In the end the leaking of his medical file proved to be a meaningless attack – other than serving as a public warning to others who might think to disturb the veil of secrecy guarding the status quo of military intelligence.

As if there was any doubt about it, the leak also makes plain that there is a spectrum of feeling, and strong feeling, within the U.S. military intelligence community surrounding the issue of disclosure. There is no unanimity. Some want more transparency while others would prefer less. Still others would regard the deliberate involvement of Congress as a device to force accountability as something deeply offensive to the authority and integrity of the military intelligence cabal. In fact this is exactly what Sean Kirkpatrick has said through a statement he published on his Twitter account.

Sean Kirkpatrick is the senior intelligence officer who in 2022, was selected to head up *AARO* (the *All-domain Anomalies Resolution Office.)* His view is that David Grusch's complaint, the determination of the Inspector General of the Intelligence Community, and the House Hearing which has resulted, amounts to nothing more than an insult to his unit.

The morning after the Congressional House Hearing Kirkpatrick tweeted the following:

"I cannot let yesterday's hearing pass without sharing how insulting it was to the officers of the Department of Defence and Intelligence Community who chose to join AARO, many with not unreasonable anxieties about the career risks this would entail." He went on to say, *"AARO has yet to find any credible evidence to support the allegations of any reverse engineering program for non-human technology."*

I am sure that Sean Kirkpatrick is quite right about the integrity of the officers under his charge at AARO, and of course it's nice that he would want to stick up for them. However that really is not the issue, as even the most casual observer could have pointed out to him. The questions brought to Congress by the means of the July hearing are really very simple. *"Was David Grusch's authorized work of investigation blocked by the units responsible for The Program, or not? And if it was blocked, was there any proper legal basis for that blocking?"*

Sean Kirkpatrick's total avoidance of those two basic questions is conspicuous and gives his scolding tweet the appearance of a misdirection. Furthermore, given the volume of information that has already been released into the public domain over the last four years concerning a full seventy years of technical inquiry by Pentagon units, the words of Kirkpatrick's statement are hard to credit. Information now available to the public includes testimony from Sean Kirkpatrick's own comrades within the intelligence community:

- Luis Elizondo, who headed the Pentagon's *Advanced Aerial Threat Identification Program* for a decade.
- Alain Juillet the former chief of French Intelligence who corroborated Elizondo's explanation of AATIP's remit with regard to retrievals and reverse engineering.
- Dr. James Lacatski, who ran the Defense Intelligence Agency's iteration of the Program, and who has acknowledged the possession and analysis of entire craft, of which David Grusch reports there are *"quite a number!"*
- Eminent astrophysicists Dr. Jacques Vallee, Dr. Eric W. Davis, and Professor Garry Nolan of Stanford University, who, on public record, have gone into some detail regarding their analysis of exotic materials retrieved by The Program.

So, Sean Kirkpatrick's claim that his unit has *"yet to find any credible evidence,"* did nothing more than illustrate the problem of compartmentalization within the intelligence community. It's ridiculous and it's embarrassing.

As the dust settles around the unfortunate Tweet, The Pentagon has announced a new authority to supervise Mr. Kirkpatrick's unit. Sean Kirkpatrick himself has been *"re-positioned."* AARO will now be directly oversighted by the most senior female officer in the Pentagon, the Deputy Secretary of Defense, Kathleen Hicks. Sean Kirkpatrick will now report to her, and she will be the unit's effective chief. It would seem the higher ups are concerned about the optics of this unseemly scuffle. Evidently the attacks on David Grusch and the claims of *"nothing to see here,"* have not played well. As Kathleen Hicks takes up the mantle of authority over AARO, she explains the change to an interviewer for *Defense Scoop*.

"I believe that transparency is a critical component of AARO's work, and I am committed to sharing AARO's discoveries with Congress and the public, consistent with our responsibility to protect critical national defense and intelligence capabilities."

At these words (with my underlinings) my thoughts return to my correspondents' insightful questions as to whether this whole exercise has really been a pretext to reshuffle fiefdoms. In November of 2023, within three months of Kathleen Hick's insertion into the schema, Sean Kirkpatrick has resigned from his directorship of AARO after less than eighteen months in the job.

"I am ready to move on," he says. *"I have accomplished everything I said I was going to do."*

As he departs Kathleen Hicks releases a statement applauding Mr Kirkpatrick's *"commitment to transparency with the United States Congress and the American public."* She says it *"leaves a legacy the department will carry forward as AARO continues its mission."*

Transparency. There's that word again. As if that were the watchword in the world of secret services. This public reshuffling in the name of transparency is a clear sign that the intelligence community, whose entire existence revolves around secrecy, is feeling the pressure on this issue of accountability surrounding The Program. As it should.

Of course, it is perfectly understandable that the research and development aspect of The Program, along with its implications about our place in the universe should be conducted under a veil of secrecy. My question is at what point do its findings become clear and important enough to belong to humanity at large? Of course there will always be special access programs needing black budgets, but from a democratic viewpoint, how much unaccounted for public money can be given to private corporations for their private research and development with no public knowledge or scrutiny?

Furthermore, which exactly are the corporate entities availing themselves of the USA's *"unaccounted for"* $9 trillion of public money, identified nearly a decade ago in a remarkable moment of economic transparency? Goodness only knows what that figure is today! What proportion of that money has been handed to the corporations to which U.S. Defense has sub-contracted all the nuts and bolts of The Program? This is public money we are talking about. For these reasons, and simply for the sake of information about The Program, I wonder why more courageous journalists, people like Ross Coulthart, haven't traveled to Virginia to sit down with the executives of corporations like Boeing, Northrop Grumman and Raytheon or to Maryland, Washington DC, or Palmdale, California to compare notes with the executives of Lockheed-Martin and Skunkworks. Surely, it's in those kinds of places we should be looking for detail of where The Program has reached.

Of course, corporate representatives are unlikely to disclose anything very interesting, being bound by layers of non-disclosure agreements and official secrets laws. But wouldn't a journalist want to sit down with them, present some salient questions and report the reaction? After all, if Brigadier-General Haim Eshed is correct, and our visitors truly have decided not to self-disclose until we understand *"what space is and what spaceships are"* isn't it important to know how far forward we are in that pursuit?

At the coalface of these scientific investigations in 2023 is a cadre of elite scientists. As I mentioned earlier, three in particular have been willing and able to meet the press and speak openly about the metamaterials they have examined as part of The Program. Most prominent are the French astrophysicist and computer scientist, Jacques Vallee, and American astrophysicist Eric W. Davis. The latter is the *"Davis"* of the *"Wilson-Davis Note"* which Richard Dolan mentioned in our conversation on *5thkind.tv* a few pages ago.

Jacques Vallee is a living legend in the field of Ufology. So much so that Steven Spielberg immortalized him in the movie *Close Encounters of Third Kind* as the character Claude Lacombe. There is even a two-character homage to Jacques in a 1996 episode of the *X-Files*. That's how revered he is. Jacques Vallee is often quoted by people wishing to debunk classic Ufology. They cite him as an eminent authority in the field who rejects the idea that UFOs are material or interstellar phenomena. However this isn't true. Early on in Jacques Vallee's journey of research into the phenomenon, yes that was his position. At that time he was more interested in our visitors' effect on the conscious experience of contactees and he was reluctant to cast our visitors as material beings from another physical region of space. Today, by contrast, along with Eric W. Davis and Garry Nolan, Jacques Vallee is paid to examine physical materials from UFO retrievals as part of The Program's efforts to reverse engineer what has been obtained.

In 2021, Jacques was filmed at his lab in Silicon Valley, speaking with my friend and film-maker James Fox. With regard to the progress of The Program at that point he said, *"The breakthrough has come with the invention of a machine that enables us to look at the atomic structure. The material [we are analyzing] is not natural. And it was not manufactured with the materials we have around us on the Earth."*

His colleague, the Stanford Microbiologist Dr Garry Nolan, elaborates on the implications of what they have found using this technology. *"I'll call it an ultra-material...Somebody is putting it together at the atomic scale. We [humans] are building our world with eighty elements. Somebody else is building the world with two hundred and fifty-three different isotopes. I intend to use this information to try to build something."*

That was what he said 2021. I wonder what has happened since. Given this information, you can see why Sean Kirkpatrick's angry *"yet to find anything"* tweet was so sorely out of kilter in 2023. As to the Wilson-Davis note, now in the public domain, we are talking about Eric Davis's minutes of a meeting in 2002, with Admiral Thomas Wilson, the chief of the *Defense Intelligence Agency*. This note provides credible, authoritative references to the intact craft, referred to by David Grusch and Dr James Lacatski, and to materials engineered off-planet, outside of Earth's atmosphere, in a zero-G environment, engineered materials which are *"not of this Earth and not made by human hands."*

The Wilson-Davis note makes public exactly what is being studied. It also shines yet more light on the issue of compartmentalization within the intelligence community because what emerges from Davis's minutes is that even while serving his country as the Chief of the *Defense Intelligence Agency*, with all his top-secret clearances, Admiral Wilson found even he was not allowed access to any further detail concerning where The Program was up to. Needless to say, this became a point of considerable annoyance to the Admiral, who regarded such information as essential to his remit. If not senior intelligence personnel like Admiral Wilson, then who exactly are the shadowy figures who do have access? Admiral Wilson's answer to Eric Davis, as recorded in the note, describes this privileged cabal as *"Corporate Types."*

Meanwhile, since 2019, Jacques Vallee, Garry Nolan and Eric Davis, have been permitted to continue sharing in broad terms the directions of their ongoing research on behalf of The Program and the word on the street regarding the retrieved metamaterials is *"spacetime manipulation."* There is, evidently, something about the physical fabric of these non-human craft which is vitally different to any material we have yet engineered on Earth. The properties of the fabric in question appear to have been artificially engineered at an atomic level, outside of Earth's atmosphere, and in such a way that the fabric itself creates a local distortion in the field of spacetime, what we might call *"a spacetime bubble."* To express it in graphic terms, if you picture the warp engines of the *Star Trek* canon, the theory of that fictional technology is that the engines of the twin nacelles somehow combine to generate an energetic field able to distort a pocket of spacetime, effectively creating a stable wormhole through which the craft can then pass. In this way the craft can cross vast interstellar distances without all the bother of actually having to travel through space. What Vallee, Davis and Nolan's hints suggest is that it is not UFO *engines* which generate wormholes and enable the craft to ping through spacetime, it is the materials from which the craft is built.

Just to remain within the canon of *Star Trek* for a moment, the franchise's more recent scriptwriters have added another layer to their concept of sub-space travel. The *Spore Drive* of *Star Trek Discovery* plays with the ideas of fusing biologics with technology and creating an intimate connection between the mind of the pilot and the technical functions of the craft. This is the cutting-edge advancement which means that the *Starship Discovery* can ping into a precise location in space, thousands of light years from its start point. This fictional fusion of technology with biologics draws my attention to an anomaly in the real world, namely, what is a Nobel nominated microbiologist doing examining the fabric of ET craft?

Why am I telling you all this? Why bother with the imaginary world of science fiction? Simply because the scriptwriters have not plucked these ideas from out of a vacuum. The concepts they are playing with can be found in the testimony of real-world U.S. naval pilots who have engaged with UFO craft which behave in exactly the way of the *Starship Discovery*. Commander David Fravor, pilots from his command, and fellow witness at the House Hearing, Ryan Graves, have described their crews' attempts to engage craft which ping into our airspace as if from nowhere and which appear to move less like a vehicle and more like a conscious entity or an animal, as if neither a molecule nor a nanosecond can be found between the intention of the pilot and the movement of the craft. These real-world observations flag the possibility of a more exotic relationship between the non-human craft and the *"non-human biologics"* of David Grusch's testimony and Richard Dolan's language of spacetime manipulation.

It is a phenomenally exciting new branch of materials science, that is until we stop and take a minute to ask what the *"corporate types,"* who would possess it, might think to do with it. I say this because I have an old-fashioned belief in democracy. So it makes me a little uneasy to think about the world's ultimate technologies, whether they should mean affordable heating for Britian's old age pensioners, or the power to ping from one galaxy to another, and to consider that these advances may be secreted far and away from any kind of democratic scrutiny, in the special access programs of corporations, whose involvement through the decades in global warfare and human misery would send a shudder down the spine of any human being. We are deep in the bowels of what Eisenhower called the Military Industrial Complex. Now I should say, for clarity, that the Admiral Wilson note did not openly name any specific corporation. The note leaves us to do the math for ourselves. The Admiral simply said, *"An aerospace technology contractor – one of the top ones in the U.S."* Perhaps I shouldn't be worried. Maybe corporate types, aerospace executives and such, are actually safer hands than our political class in the twenty-first century and, to be fair to the *"corporate types,"* I actually don't find that too hard to believe. Indeed, as Richard Dolan reminded us earlier, our politicians are often so far removed from the real power in the corporation-dominated realpolitik of our time, that a focus on parliamentary activity might prove to be something of a misdirection.

In that vein, upon their respective retirements, Presidents Woodrow Wilson and Dwight Eisenhower both warned the American people, and indeed the listening world, of the dangers of geopolitical power shifting from our parliaments to an unaccountable web of military, industrial and other corporations.

I think it's relevant to mention here that Dwight Eisenhower was not a son of corporate America. His family roots lay in soil a long way from the realms of corporate life, warfare and power politics. His heritage lay in the humble and hard-working counterculture of Mennonite Christianity, which is a tradition far removed from the acquisitiveness, privilege and classism of America's elites. It was a family culture rooted in historic decisions to leave behind the comforts of conformity and established wealth in seventeenth century Europe. Specifically to avoid the militarization of their society, the Mennonite pioneers made their epic migration to America, working together to create a better society in the new world. That was the culture which gave birth to this five-star general.

Eisenhower was one of a passing generation of western leaders who occupied their seats of authority not because they were career-politicians with a thirst for power, but because they sincerely believed that, with wise and enterprising leadership, we can build a better world. Maybe General Eisenhower was the last president of that kind – perhaps with one exception, maybe two. And was he, I wonder, the last American President to be allowed a seat at the alleged intergalactic table, in person, on a need-to-know basis, as calculations were fine-tuned as to our most useful alliances as a soon-to-be-space-faring civilization? If so, we can only imagine what must have passed through his mind as he was read in to a new world of exopolitics. How could a man of the 1950's wrap his mind around every moving part in that equation? How could anybody?

If the story of the 1954 meeting is true then it reflects, if not an invasion, or a colonization, at the very least an annexation of some kind. What, I wonder, would the addition a new uber-government mean? Is it just a matter of oversight and supervision? Are we the objects of what is no more than a monitoring operation, with our observers waiting patiently in the wings for us to reach the techological bar? If it is more than that and we have covertly been annexed, how would the general public discern it? What would be the signs you and I might expect to observe at ground level?

The stories I hear from Kam in Hawaii, from ancient Babylon, Greece, Mesoamerica, and in the pages of the Bible, reveal on the one hand the prospect of rapid technological and social development. Think Asherah, Hun Hunahpu, Prometheus, Mbab Mwane Waresa, Oannes and the Apkallu. Have we seen such an accelaration of technological progress since 1954? I would say so.

On the other hand we should consider the shadow side of the Bible's elohim, the Norse Aesir, Babylon's Apkallu, the Anunna of Sumer, and Kam's stories of the Anunu of Hawaii. Surveying

183

the last seventy years it isn't hard to discern a progressive machinisation of human society and an escalation of means by which the movements and activities of large populations can be monitored, managed and controlled. Looking back over the last seven decades, haven't we seen a phenomenal acceleration in that direction too?

Are these shifts purely the epiphenomena of human progress or are they the signs that we have long since been invaded and that forces far beyond humanity have had a hand in our ongoing story for a long time? Would you and I recognize the signs?

CHAPTER THIRTEEN

The Currency of Conquerors

Molokai, Hawaii - 2024

How would you and I know if a covert invasion had already happened? What did our ancestors think it looked like? What signs did they think you and I should look out for? Being a hermeneutics guy, it is easy for me to find correlations between the hidden aspects of the Biblical narratives and Kam's explanation of the Mo'olele. In their overlap I find an eyewitness vantage on these questions. Kam's stories are subtle and multi-layered, but the broad brush is unambiguous. His ancestral memory is of a flesh and blood invasion.

Because of the dragon language of the Mo'olele it is all too easy for the western mind to dismiss this ancient canon as mere fable and fairy tale. However, within these narratives I have already identified an insightful and nuanced understanding of social history.

Listened to with an open ear, the Mo'olele casts a surprisingly socio-political light on the question of covert takeovers and annexations. The correlations shared by the Mo'olele and a paleocontact reading of the Bible are striking and the many parallels in other ancestral narratives around the world add to the sense of a coherent picture on a global canvas.

Whether we sit at the feet of elders of ancient Nigeria, Cameroon, Mali, Egypt, Sumeria, the Levant, India, Iceland, or ancient Greece, we will hear the same story of beginnings as it relates to human governance. It says that prior to the familiar bloodlines of kings and queens and successions of presidents and prime ministers, our ancestors were governed for a long time by non-human others who were so advanced that their power seemed unassailable, until the time came when they relinquished their authority and handed it over to human leaders. This is the schema of the airborne Kings of the Vedas, the Aesir of the Vikings, the Elohim of the Bible, the Ogiso of Benin, Abassi and Atai of Nigeria, the Nommos of Mali, the Anunna of Sumeria and the Anunu of Hawaii. In every case these beings arrive, colonize, delegate and leave.

We could liken this pattern of arrival, exploitation and departure to the way we invade each other's countries. When we arrive it must be with a show of force, fleets of ships, guns blazing. We overwhelm the locals with our superior firepower. We become the law, the police, the newsagency, the education department, and the army. Our right to rule is established through our unassailable physical presence. Once established, once we have changed the currency, set up the banks and exchange rate mechanisms, once we have set in place the trade deals and the commodity prices, we can invite the locals to populate the middle tier of public service. Locals can become teachers, police officers, priests, judges and ministers.

Eventually we can return home, taking our ships and our guns with us, secure in the knowledge that the diamonds and gold and cocoa and sugar will still be a source of profit for us. Even from afar we can continue to enjoy the benefits of sitting at the top of the economic tree, shaping international relations, secure in the knowledge that we have a military staging post far from home should we ever need it for strategic purposes. For all these reasons, we hand over the visible reigns of power to local successors, sometimes at the time of our choosing, sometimes a little sooner. And this is exactly what we hear in the annals of the cultures I have just named. This is how we saw advanced others do invasion. This is how we do it.

Kam is totally upfront and straightforward in his language of invasion. He speaks simply and clearly of *"When the Mo'o came,"* or *"When the Ahumanu arrived..."* or, *"When the Anunu took over..."* Though most believers read the Bible as metaphor and God-story, the root meanings of key words carry this same story of physical invasion and conquest.

In the Hebrew scriptures the visitors are remembered as the *Seba Hassamayim* the *Airborne Army* or *Sky Army,* and in Kam's narrative too the Mo'o, Anunu and Ahumanu arrive with guns blazing. The Bible's Yahweh or El Shaddai is presented as a physical and militaristic being. In the book of Job, his physical attributes are described through comparisons with other biological entities, and his flight technology and weaponry are described in detail in the books of Psalms, Exodus and Ezekiel. Both the Mo'olele and the Seba Hassamayim narratives of the Bible specify physical weaponry of phenomenal destructive power and these aspects repeat in the Greek stories of Zeus, and the Indian stories of Shiva.

However not all physical arrivals are the same and not all are violent. As we saw before, benevolent interventions which taught our ancestors to live in balance with the land appear in the ancestral stories of First Nation Americans and Aboriginal Australians. The pattern repeats in the Asherah stories of the Levant and surrounding regions, the Tagalog story of the Philippines, the indigenous Maya of the Yucatan peninsula, and indigenous Amazonian story in northeast Brazil. All these cultures describe the physical presence of advanced non-human beings who helped their ancestors make the great leap forward into sustainable, combination and rotational farming. Some of these stories hint that these ancient tutors may have been resident on the planet for a long time prior to their assistance of homo sapiens. This would include the Yolngu and Mohawk stories. Other narratives are specific as to the cosmic origin of these ancient tutors, with oral and archeological artefacts of the Cherokee and Asherah narratives, for instance, both identifying the Pleiades.

The various thematic clusters within this great canon illustrates that we may have been visited numerous times throughout our development as a species by various civilizations with diverse cultures and differing agendas, and it is possible that fragments of memory relating to many kinds of intervention have been woven together in our indigenous stories of beginnings.

We saw this in Genesis, where a sequence of stories of beginnings and resets has been pasted together to form what, on first inspection, looks like a single narrative. Kam's story seems pretty linear on first hearing, but it too may be a story of many layers with some elements more ancient than others. Nevertheless, he is eager to impress upon me that the cultural shifts which followed the invasion of the Mo'o were palpable and not positive. So let me run through what those shifts were in essence, and how they echo in the mythology I know best, that of the Bible.

Draconian Government

Draconian governance means a top-down social order run on the basis that the governors are served by those who are governed – the very opposite of the democratic ideal. This dynamic of a servile population is reflected in the Mayan story of origins enshrined in the Popol Vuh. In that telling, the dragons (*Feathered Serpents*) say to one another, *"Let us make avatars to do the unpleasant work for us and bring us our food."* It is a social order founded on the threat of violent consequences.

This is a story whose invention would serve nobody's interests as it flatters neither our ancestors nor their ancient overlords. The same order is there in the Hebrew scriptures in which Yahweh and other advanced beings, which are described as dragons or feathered reptilians, each organize their own human colonies, employing the humans to keep them supplied with beef, lamb, gold and virgin girls along with the first fruits of every harvest. Every idea of sacrificial religion is founded on this ethic of human servility. Arguably, it undergirds the class system and social order of cultures all around the world, irrespective of any religious association.

In western society at large, many of our social and economic structures appear to be based on this fundamental idea that power and wealth should flow from those at the bottom of the pile to those at the top, despite centuries of attempts to democratize our societies. So many of our patterns of employment, taxation, and governance are built on the dynamic of the *"lower downs"* serving the *"higher ups."* Any individual or group which offends or fails to comply with this basic order will be met by the force which holds this order together, namely the strength of the powerful against the powerless.

On a societal level, for instance, if you or I refuse to pay our taxes or keep the laws imposed by the higher ups on the lower downs, then before long our property will be invaded, we will be physically overpowered, manhandled, and taken by force to be locked in a cage until our *"debt"* to the higher ups is paid. To maintain control of its humans, any dragon must constantly advertise the threat of dispossession, death or a dungeon to remind the humans why meek compliance to the dragon's wishes is really the easier way. Same as it ever was! This is an unattractive aspect of how social cohesion is maintained in our own societies, and we might not choose to look at it very often, but to this day this is the threat which lurks behind the façade of democracy and civilization. It is the prospect of dispossession, death and dungeons that keeps our societies in order.

The Manipulation of Currency

Whether following the Hebrew narrative or listening to Kam recount the Hawaiian Mo'olele, the dynamics of draconian rule are the same: invasion followed by the trickle up of resources, the ordering of society through threats and violence, and the incremental dispossession of ordinary human beings through sacrificial taxes, tithes and tribute.

"The moment we agreed with the value of their tokens which they called money, from that moment on we lost our freedom. That was really how they enslaved us."

This is how Kam describes his ancestors' unwitting surrender of their sovereignty. This acceptance of the colonizers' money was not a mutually beneficial arrangement. It would be the colonizers' privilege not only to control the supply of the tokens but also to define and redefine the value of them. People might work for the same number of hours, for the same number of tokens, from one year to the next, but if the value of the tokens is depreciated, then those issuing the tokens can increase their power over the workforce without needing to raise so much as a bow and arrow.

Is that so different from how we order our society today? We too have to work for money without any control over the value of that money. It is depreciated every single year that we work for it. Hence, where I live in Australia, in the 1960's a working father could buy a house for his family, provide holidays, support his wife at home with the children, pay for their college education, and expect to retire in comfort, all on the basis of a single income, his own income, on an average wage.

In 1966, in Australia, the average house price was 1.6 times the average household annual income. This was at a time when most households existed on a single income. What that meant was if a married couple were to work full-time on average incomes in an average house, allowing for the fact that at that time women were paid significantly less than men, they could own that home outright within two to three years. Try doing that today.

In 2023 in Australia the average house price was 7.6 times the average household annual income, bearing in mind that two thirds of those households are dual income. Roughly speaking, the average household today has to make thirteen times the financial commitment to buy a house compared to 1966, the year when property ownership peaked in Australia. What has changed is not the number of hours in the week, nor the work ethic of the country's citizens. It is the value of the currency that has changed.

In the gospel of Matthew, Jesus is asked about the ethics of taxation. In reply he makes a profound point about the meaning of money in the international financial system of the Pax Romana. For context, Rome had conquered and annexed Palestine, and then imposed the international imperial currency of libri, soldi and denarii. At some level the people would now have to work for Roman money, because it was in Roman money that they would have to pay their taxes. Following the same pattern as the Mo'o of the Mo'olele, the invaders first issued the tokens and then they asked for the tokens back. So when put on the spot regarding the ethics of that, Jesus neatly answered, *"Whose money do you think this is? Take a look at it! Whose face is stamped on it?"*

There was no ambiguity to it. It was the conqueror's currency, minted for the conqueror's benefit. Jesus' question about who the money really belongs to remains relevant in the twenty-first century. Around the world today, there is a significant grassroots move afoot, as more and more people are electing to hold their monetary wealth in currencies not controlled by governments and banking cartels. When wealth is held in other forms, governments and banks can no longer control or alter its value. This is the key to the crypto currency alternative. As this trend continues, I note that some of the banks in Australia are now refusing to give customers their own money if those customers wish to use it to buy crypto currency. If I go to my branch, for instance, and ask to withdraw any cash from my own account, for my own personal use, the teller will now ask: *"What is the money for? We are required to ask!"* Jesus' question is as relevant as ever: Whose money do we imagine it is?

Neither was the Jesus of the gospels any more impressed with the local taxation system, the Jerusalem Temple-tax, to be paid in the local currency issued by the Jerusalem Temple. Once again, the temple issued the currency and then received it back as tax, tithes and payments for priestly services. To paraphrase, Jesus asks the people, *"Who do you think pays this tax? The elite? The children of the elite? Of course not? The elite and their children are exempt. No, these taxes are for the masses to pay and for the elite to receive."* (This all plays out in *Matthew 17 & 22*)

Both the Roman currency, and the Jerusalem Temple currency were in essence mechanisms to separate the grassroots from a proportion of their wealth and reassign it to the higher ups. The polarization of interests, which Jesus highlights, brings us to another feature of the ancient narratives, with an idea of what life may have been like when we were ruled over by non-human governors, who naturally had no sense of fellow-feeling with the mass of humanity.

Proxy Wars

Whether we read the sagas of the Kings in the Vedas, or the elohim in the Bible, we are reading a litany of wars in which the overlords conflict with one another over resources and hegemony, and in which the human beings, who have no argument against one another and no real stake in the outcome, become the foot soldiers and collateral damage, deployed and dispensed with by the powers with next to no sense of grief at the human cost of their squabbles

The most famous illustration of this divorce of agendas in the politics of war has to be the famous *Christmas Truce* of the First World War, a war which over four years slaughtered nearly ten million infantry. The Christmas Truce happened at the end of the first year of the Great War in 1914. On Christmas Eve of that year, the boys manning the British trenches realized that they could hear singing rising up from the boys in the German trenches. They were singing Christmas carols. British boys then started shouting across no man's land with a greeting of *"Happy Christmas!"* And shouts of *"Frohe Weinachten"* came in reply. In the hours that followed German and allied infantry began mixing in no-man's land, burying their dead, exchanging gifts, and playing soccer together. Why? Because the ordinary, working-class boys of Britain had no argument with the ordinary, working-class boys of Germany. They actually did not want to kill each other. They wanted to play soccer and celebrate Christmas.

The argument was at an elite geopolitical level, an argument among royal cousins over which one of them owned and governed which lands. It was the ruling elites breaching their mutual agreements, whose actions and reactions led ultimately to war, financially supported by powerful banking families who managed to profit from the carnage by bankrolling both sides of the war. Accordingly, as the record shows, it was the higher-ups in the great chain of command of WWI who took emphatic action to ensure that peace would never be allowed to break out on the front lines like this ever again. Peace would be reached only when the elites were ready to agree new terms, not whenever the grassroots might want it, even if that delay should cost another twenty million human lives. Which it did.

In all the years since, the worst atrocities have been committed by leaders who do not want peace. They do not want harmony breaking out between their own people and the neighboring people. They welcome the cycle of attacks and reprisals. The people whose homes, schools, hospitals and supply lines are bombed certainly do not welcome it. These are not grassroots wars. They are wars among the powers, with innocent citizens deployed and caught in the crossfire. It is precisely the same scenario as is described in the narratives of the Mo'o, Anunna, Kings, Aesir and Elohim.

If ancestral chronicles of proxy wars initiated by non-human overlords weren't sufficient warning, the authors of the gnostic texts of early Christianity offer us another lens by which to interpret the politics of war and recognize a non-human hand at play. The gnostic texts survived the purges and book burnings sparked by the Roman Emperor Theodosius in the 4th century CE only by being buried in the Nag Hammadi desert for their preservation. These incredible scrolls and parchments warn us about beings called *archons*. Archons can be described as parasitic, non-human, energy-based entities which feed off the heavier emotional energy of organic life. Like a virus they intrude on the health of human beings, arousing anxiety, paranoia and aggression, leading to negative behaviours and potentially a cycle of aggression, fear and despair upon which the archons can then feed. Accordingly the favourite sport among archons was manipulating human leaders into warfare.

Whether we take the archon stories as representing non-human entities in the real world, or whether we take them as figurative of psychological patterns, the take-home message is the same: Pay attention to your emotional state, lest you be manipulated into bad decisions, with bad consequences, which will impact on your wellbeing and the wellbeing of others. To be clear though, the gnostic writers took these stories as warnings about genuine contact experiences of a malevolent kind.

This concept of cloaked, predatory, parasitic company, manipulating political leaders into unnecessary conflicts, can also be found prior to the gnostic period, the first three centuries of Christian thought, in the earlier canonical writings of the Hebrew scriptures. The writer of *I Kings* records a moment in which Micaiah, a man known for a shamanic ability in far sight, succeeds in remote viewing the machinations of the *El-Ba'adat*. There he witnesses a conversation in which the non-human parties sitting in council agree to manipulate one particular nation into invading another country on the basis of false intelligence. Clearly the Hebrew scribe who carefully penned that story, hoped that future readers would be alert enough to recognize any threat of such a manipulation happening in their own generation. It's a story that deserves a higher profile in the present day.

In one way or another our ancestral stories give us accounts of the past to help us discern the same kinds of manipulations they suffered in their own day. *"These are the signatures of archonic manipulation,"* they would say. *"These are the signs of El-Ba'adat, Secret Commonwealth or draconic influence in your governments and royal courts."*

In the following chapter we will explore some more signatures and signs of covert non-human influence as we survey the world around us today. Come with me and take note if any of them should ring a bell.

CHAPTER FOURTEEN

Stressors and Soundbites

Molokai, Hawaii - 2024

By definition, draconian governance is a reign of terror, and the wave of dragon story that dates from ten thousand years ago, is full of it. In that second wave of the world's canon of dragon story, from Germany to Egypt, from Georgia to Babylon, the narratives are shot through with the central theme of social order maintained by the threat of violence. It sounds like another time and another world, like something from Terry Pratchett's Discworld or Tolkien's Middle Earth. Yet on closer inspection the details and dynamics of these narratives are strangely recognizable, and not least in the draconian use of terror as a tool for social control.

Governance by Threat

Of course a flesh and blood dragon can terrorize its citizens overtly and without shame. The periodic public immolation of any unruly citizen can serve as a regular reminder to the general population as to who's in charge and why. By contrast, democratically elected governments must present themselves as servants of the people. Public servants are not supposed to rule over the people, let alone terrorize them, at least not in theory. Accordingly, elected powers have to be less direct in any use of physical violence. Irresistible force has to be applied one degree removed from government authority It must be done by a separate agency, such as an independent police, the army, or private security companies, such as the ones which operate our prisons, patrol our shopping malls, and run detention centres for troubled youth or asylum-seekers.

In this way, state-endorsed violence can be meted out, without state accountability towards any group – but most probably one that is either unlikely or unable to vote for the sitting government. The target may be an ethnic group, indigenous people for instance. It may be a social group, protesters of any kind, gypsies, asylum-seekers, street people, orphans, convicts, or people with health problems through drug addiction. First, official voices must objectify the groups in question so that those people come to be seen as something other than an integrated part of society in general. Next officials must vilify the chosen group without any hint of understanding or empathy. Once the public has become habituated to this more brutal and vulgar tone from senior officials, the government can then begin doing other things without fear of popular complaint, things like introducing overbearing laws or allowing light to be shone on miscarriages of justice, or police brutality or deaths in custody, while conspicuously taking zero action.

These measures have nothing to do with responding democratically to the groups being victimized. Quite the reverse. The message is for the public. It is a stern reminder of the power the state can assert over the individual should it see fit. And this undercurrent of threat is the very essence of draconian government. So, if you see an investigative journalist persecuted by a government to the point of insanity, or a cancelled to the point where they are no longer allowed access to the media, or to speak publicly, or sell books, make a living or retain their property, make no mistake the message is for the general public. The one who should take warning is you.

Similarly when refugees are treated inhumanly, are denied medication or legal representation, once again the message is not for terrified and traumatized families, widows and orphans, leaving everything behind and risking everything to flee for their lives from bombs, civil war or tyranny overseas. Those people are not sitting at home watching your news programs before they make their desperate decision. The message is for the viewer. It is for you.

Where in the twenty-first century we use words like *journalist* or *protester*, our Hebrew ancestors might have used the word *haro'eh* (seer) or *nabi* (prophet.) Time and again through the annals of the Hebrew scriptures, any *nabi* or *haro'eh* whose public messages departed from the accepted script would find themselves made an example of, being rounded up, slapped in solitary confinement, buried in holes in the ground, or beheaded – a sober reminder for all others as to the dangers of independent thought or reportage.

The psychological foundation for this kind of governance by threat is a general suppression of compassion and empathy, a dumbing down of the emotional intelligence by which we imagine our way into each other's shoes, to get a sense of how things look and feel where each of us is standing. A vision of society that sees *he* as *me* or *she* as *we* is the bedrock of social harmony and solidarity. Governance by threat pulls a different lever. Draconian government thinks differently whereby state violence against *"she"* must make *"me"* feel threatened. If I am shown how brutally *"he"* was treated *"me"* must be made to fear that *"we"* may be treated in exactly the same way, or even worse. Social harmony is the enemy of tyrannical government, which will always prefer an insecure, needy and atomized populace for ease of governance. It is the modus operandi of management by threat.

In his garden on the island of Molokai, Kam told us that the Mo'o governed by turning people into rivals and competitors. Similarly, back in the seventeenth century Scotland, Robert Kirk in Aberfoyle argued that a pervasive lack of compassion is perhaps

199

the strongest indication of non-human players influencing the elites who direct our covert layers of government.

Governance by Econometrics

This kind of callousness can manifest in ways that have nothing to do with terror or violence at all, but which ultimately are no less dehumanizing. In the present-day callous governance can fly under a different banner, something innocent-sounding, something like *econometrics*. Let me take you on a brief tour of public policy in my own part of the world to show you what I mean.

Firstly though, let me clarify how I am using this word, *econometrics*. I am talking about how we understand the economy, and in particular the way in which we have become accustomed to measuring a nation's economy in dispassionate, value-free, mathematical terms. This approach tends to lead our policy-making into a kind of twilight zone, in which the strength of a nation's *"economy"* is understood as having nothing to do with the actual material wellbeing of the nation's people.

For instance, consider the economic soundbites to which we are treated on a daily basis at the end of the national news. We are told about exchange rates, the Dow Jones average and the FTSE 100 Index, as if these figures offer the viewer a snapshot of the fortunes of the nation on that given day. But what do these numbers really mean to the average viewer? As for the few who really understand these figures, does anyone suppose that those people rely on the TV news to inform their investment decisions from day to day? Of course not. These mathematical soundbites are the continual assertion of a vision of the economy as a pure matrix of money, interest and mathematics. Human beings do not figure in this picture. Why would they? Human beings? They're the concern of anthropology or social studies, not economics!

200

But why not, instead, conclude the daily news with other metrics? We could instead feature the daily count of families with two working parents consigned to living in cars or trailers because Hawaii is not the only place where that happens. Why not include a daily log of the number of people incapacitated because they can't afford the available treatments? In Australia the federal bureau of statistics keeps a detailed record of how many people each month give up looking for work in despair, stop claiming benefits and go into financial freefall. We have the numbers. Why not broadcast them?

On a lighter note, hopefully, we could report the volume of today's crop yields. One metric relevant to everybody in the country would be the average ratio of household income to the cost of living. All kinds of metrics could be chosen if the purpose is to give a snapshot of the nation's fortunes. But no, as long as the exchange rates are where they need to be and the Dow Jones average is buoyant, then we can sleep easy knowing that the *"economy"* is strong. Your and my inclusion in that economy is really not relevant. Yes, the nation may be blighted by a funk of sickness, homelessness and despair, but thank God the economy is healthy!

This radical divorce of human economics from *"the economy"* raises an obvious question, *"Whose economy do these metrics actually describe?"* Or to put it another way, *"Whose interests are served when we frame the economy in this purely mathematical way?"* It is a paradigm which guides us to celebrate gains for big corporations as if they are gains for all of us, without worrying too much about any associated social costs. After all what are *"social costs?"* Where are they in the science of measuring the economy?

The vocabulary of *"social costs"* is really value-based language, whereas econometrics speaks the pure dispassionate language of mathematics. As the eighteenth-century master of economic

201

science Adam Smith pointed out in his book *"On Moral Sentiments,"* untrammeled economic forces are amoral. Hence their metrics are purely dispassionate and factual.

We can see this disinterest in human life on full display when governments legislate for access to safe, clean drinking water to be denied to ordinary human beings and given instead to industrial manufacturing corporations. This very thing is happening today in South America, North America and where I live in Australia. The same disinterest is in evidence when governments alter regulations with the effect of increasing profit margins for corporate food producers while simultaneously toxifying the food supply for the general public, thereby reducing quality of life and life expectancy. Hence, chemicals which are legally banned from food for human beings in Europe and Australia because of their high toxicity are allowed to freely pollute the foods sold to Americans. Apparently, it's alright for the citizens of America to ingest them. Why? Because the econometrics are good. It's just a shame about the health implications!

Again, one has to ask: *"To whom or to what do these human costs not matter?"* What is at the heart of this lack of human fellow-feeling? Is it purely a matter of human selfishness? Or is there a non-human explanation? Our Chinese ancestors would say that this is typical of the Jade Emperor. Ancient Sumerians would tell you that Enlil was wired this way. The scribes of the Bible would say that Yahweh was no different, and Kam would tell you the Ahumanu were just the same. These figures had no fellow-feeling with human beings simply because they weren't human. But how would we answer today?

This morning, on a break between paragraphs, I was treated to a startlingly honest outburst from the chief economist for one of Australia's major banks during an appearance on ABC news. Randomly, I happened to catch his turn of phrase while I was making myself a coffee. Without any hint of embarrassment the bank's economist told the reporter,

"We want enough Australians to be out of work so as to optimize the money markets and mitigate against inflation."

Is that right?

Take a moment to ask yourself who the *"we"* is in that sentence, the *"we"* who want *"enough Australians to be out of work!"* Years ago, unemployment was regarded as a scourge. It was understood as a tragedy for families and a social ill. Today, apparently, it is a useful device by which *"we"* can *"optimize the money markets."* Once again, it's a case of econometrics *versus* humanity, with humanity coming in a poor second.

I love my country, Australia, and I count it a privilege to live here. I'm going to give you another antipodean example only because it's the place I know best. Australia constitutes the world's twelfth largest national economy. Yet if you travel no more than thirty minutes from commuter-train terminals which service the state capitals, you will find yourself driving through towns which were positively thriving a mere generation ago but which today are, apparently, no longer viable. Look out the window and you will see that you are passing by what used to be banks, post offices, local stores, schools, medical practices and local hospitals but which are now either residentialised, abandoned or demolished. All that business, essential to the life of a local community, has been withdrawn from the landscape, only minutes from the state capitals.

How come? Has *"the economy"* shrunk? Not by any means. In fact, from 1980 to 2022 Australia's gross domestic product per capita multiplied by a factor of 6.5. In 1980 it was 150 billion dollars US for a population of fifteen million. By 2022 it had grown to 1.7 trillion dollars U.S. per twenty-six million. The economy has mushroomed. Yet somehow this phenomenal growth seems to evaporate no more than fifty minutes out of the state capitals. It is certainly not that everywhere is a story of decline. Far from it. In that same period our cities have boomed and transformed. Yet fifty minutes away previously thriving local communities have morphed into ghost towns or semi-populated dormitory-ville.

This extraction of economy from beyond the major cities is not the outcome of some partisan policy. Nobody voted for these changed priorities. The inexorable death of these communities, so close to our state capitals, reveals an uber-policy which has played out over the decades, transcending a whole succession of democratically elected governments of various political colours. This long-term overarching policy has been decided higher up the food chain, and like the land enclosures of past centuries, it has forced people off family properties in the remoter regions, out of the regional towns and districts, into the major cities and into ever denser forms of housing. Econometrics is the reason.

This little repeat of history might not sound too worrying if your livelihood is in the city. But what if it isn't? For instance, in March 2015, the then Australian prime minister Tony Abott met the press to back the Western Australian government's decision to *"de-fund"* one hundred and fifty Australian regional communities. This *"de-funding"* meant the withdrawal of any government commitment to maintaining infrastructure, industry support, essential services, health or education from the communities his government had just earmarked. Accordingly, any parents living in these communities would either have to relocate or fracture their family and send their children away to be housed with other families in order to be educated in the cities. The one thing every last one of these fifty communities had in common with each other was that they were all predominantly indigenous communities, home to Aboriginal Australians, who to date have occupied their particular parcels of country for tens of thousands of years. Never mind that! Henceforward, any decision these Aboriginal Australians make to remain on their traditional lands is, so said the prime minister, a *"lifestyle choice"* that the Australian government will no longer support. Just process that for a moment. The Australian government, state and federal, will no longer support Aboriginal Australians remaining on their traditional land. And this is the twenty-first century.

When challenged on this policy, Mr. Abbott doubled down on his argument for this fragmentation of Aboriginal families by the state on the grounds that this will support the children's *"full participation in Australian life."*

The elohim of Biblical memory, the ancient Mayan rulers, and the Ahumanu of Kam's Mo'olele all exploited the device of forcing their human subjects to surrender their children, knowing that this was a modality to break people's spirit. This narrative background makes me reflect on the true emotional impact of policies that require indigenous families to send their children away for them to be *"absorbed into our mainstream."*

It is possible that Mr. Abbott himself had not imagined his way into these families' homes to truly understand the impact of the Western Australian policy, but whatever the intentions of the politicians involved, to enforce such a policy is in reality a profound psychological assault. Yet in the twenty-first century these policies are presented as simple economics and pure maths. Mr. Abbott didn't put his argument forward in the emotive vocabulary of racism, nor in the twentieth century language of needing to *"assimilate the Aborigine,"* nor in his predecessor John Howard's language of indigenous Australians needing to be *"absorbed into our mainstream."* No, purely from an econometric viewpoint, he insisted, this new policy was really unarguable. It was a matter of basic maths and, as such, couldn't be argued with. Two plus two is never five. And that is how the policy was presented.

Now, of course no alien influence is necessary to produce callous and inhuman policies of this kind. Human leaders are perfectly capable of doing that without any need of assistance from the Jade Emperor, Enlil, the Ahumanu or any other ET with ambitions for project Earth. However, if our ancestors were to hear policy announcements of this kind, they would instantly recognize the hymn sheet from which these songs are sung. They wouldn't hesitate to inform us that these are the marks of *draconian* leadership, leadership modelled on the cold-blooded dragons of ancient memory.

Kam explains to me that when the Anunu, the Mo'o and the Ahumanu first came to Molokai, they taught his ancestors to change their way of thinking about their life as a society. Greed, division and exploitation were both modelled and taught by the Anunu. In an environment of engineered scarcity, self-ism rather than fellow-feeling were the watchwords of the new regime. The Anunu inculcated it into their humans because an atomized and selfish population was, so they reckoned, far easier to manage than populations who knew how to think and act as a community. As I listen to Kam, I wonder if his ancestors had enshrined this story within their cultural canon as a way to help their descendants recognize the clues of Ahumanu annexation should it ever occur again.

Stress Arousal

One of the first discussions held by the El-Ba'adat in the Hebrew scriptures concerns the restriction of people's access to food and water and life-saving medications. This is part of the Genesis 3 narrative. This scheme of artificial scarcity was introduced as a way to mitigate against a more conscious and intelligent population. Generations of translation choices have almost completely obscured this layer of the story. However if I read this passage in Genesis 3 alongside its Mesopotamian source narratives in the Enuma Elish and the Epic of Gilgamesh, the anti-human dynamic of it becomes unmissable. It is a model of engineered scarcity as a tool for social control.

This scenario in which human stress is deliberately heightened for the sake of easier population management echoes precisely in the Mayan narrative of the Popol Vuh. In that narrative, the Feathered Serpents' chief genetic engineer, Kukulkan, is tasked with maintaining human health and intelligence at a manageable level. When the Feathered Serpents realize that they are struggling to manage the recently engineered homo sapiens they look to Kukulkan for a solution. After a sequence of long nights in the lab, Kukulkan engineers a vapour which when sprayed over human populations will brain damage them, reducing the level of their cognitive powers, powers which hitherto have proven too advanced for the feathered serpents to be able to manipulate. The toxins in the vapour will put that right.

The story reappears in Nigeria in the origins story told by the Efik people. It speaks of advanced non-human beings called Abassi and Atai who, having engineered human beings, recognize that they are struggling to control human civilization as it grows and progresses. Abassi and Atai's response is almost identical to Kukulkan's. They release devices into the environment which damage the health and mental health of human beings, making the people more stressed, more aggressive and shorter lived.

Later in the Biblical story, Yahweh, like his fellow elohim, imposes patterns of austerity in a time of surpluses. He does this by requiring the human beings living on his lands to surrender taxes to his proxy kings, and tithes to his priests, and to surrender all their first fruits to him, which means the cream of the crop of every harvest. Furthermore Yahweh's strict requirements as to which animals will be deemed acceptable for sacrifice ensures that the farmers can no longer use the best and healthiest specimens of their livestock for breeding. No, these are the animals which must be slaughtered and burned as sacrifices for Yahweh. (*Leviticus 1, 5, 12, 14, 27.*) In this way, even in the most prosperous of periods, the stress and uncertainty of living on the land has been raised, totally unnecessarily, and by several bars.

Today, intentional stress arousal as a tool for governance can be gauged in a whole range of political phenomena such as:

- Arbitrary austerity policies imposed in fabulously wealthy countries.
- Aggressive policing and heightened security measures, which unnecessarily harass and intimidate law-abiding members of the public.
- Deregulation of highly profitable corporations, permitting them to pay the lower tiers of their employees' wages so low that they cannot live without taking other jobs or seeking government support.
- The constant broadcasting of irrelevant yet traumatizing news, without any follow-up or resolution, sounding a continual and unbroken message of danger and insecurity.
- The toleration of *laissez faire* or *market economics* which allow the decreasing affordability of life's essentials to become not the concern of government.
- The refusal of governments to legislate to ensure the public's access to human essentials such as safe drinking water, essential utilities, secure and clean food supplies and vital medications.

These are all things which could be managed differently for the sake of human health and wellbeing if the agenda of our higher ups were to preside over a happy and thriving human population. Of course none of these measures are the monopoly of Elohim, Anunna, Mo'o, Ahumanu or any other version of a non-human or extraterrestrial power. They are familiar devices for the management of large populations. Just because they are at the cynical end of the spectrum doesn't make them extraterrestrial. However, what catches my attention is that there is so much agreement among the world's canon of indigenous narratives to indicate that our ancestors saw these as the dynamics of non-human interference in antiquity. Our ancestors regarded these kinds of cultural shifts as the hallmark of ET interference and annexation. Hence, we can say that these are the kinds of signs our ancestors directed us to look for in order to discern any unwelcome, non-human influence in the geopolitics of our own day.

Now, if this was where the story ended it would be a grim picture indeed. Didn't our ancestral elders offer us any kind of antidote? Didn't they leave behind any clues as to how to take a step back from these subtle and not so subtle aggressions and find a way forward?

As I have pondered these questions for myself, I have turned to a little gem from my theological background, a nugget of powerful wisdom kept from my travels through the world of eastern orthodoxy. Between the lower Danube River and the Black Sea, in territory shared by Romania and Bulgaria, there lived a hermit in the fourth century of the common era called John Cassian. He was a master of Kam's *open manawa*, a believer in accessing higher cognitive abilities and a practitioner in downloading cosmic *na'au*. He was that kind of hermit.

John Cassian's recommended response to emotional, physical and psychological assaults was always to maintain personal sovereignty by moving in the opposite spirit. His catchphrase was *"Contaria Contrariis Curantur,"* which roughly means, *"Opposites are healed by opposites."* His notion was that people with empathy and compassion cannot be manipulated by the opposite spirit. Xenophobic propaganda or demonizing portrayals of other countries simply won't work on them. Remember that it only took the sound of German boys singing Christmas Carols in the trenches 1914 to remind the boys in the British trenches that they had no argument with one another and that they actually did not want to kill each other.

Cassian taught that when served with a diet of fear and hostility we should respond with choices rooted in love and understanding. When served with a counsel of despair we should respond with a story of hope and inspiration. If afflicted with bullying and harassment, we should respond with patience and empathy. Ultimately, when our love and solidarity with one another exceeds our fear of dungeons or dragons, that's when we become able to confront our dragons and let them know that their era of terror is finally at an end. It is the application of the old adage that *"love conquers fear"* which empowers the people to overcome the atomizing agenda of any draconian regime.

The example of Gandhi's passive resistance to Britain's annexation of India, the victories of the Civil Rights movement in the U.S.A. in the 1950s and 60s, the fruition of Nelson Mandela's subtle negotiations from his prison cells on Robben Island, Pollsmoor and Victor Verster, all demonstrate that John Cassian's advice to move in the opposite spirit has real power, whether applied individually in the workplace, or collectively in the public square. When those who govern us face a public that is less fearful and more courageous, less atomized and more united, less paranoid and more empathic, less xenophobic and more compassionate, less selfish and more loving, then their appeal to the public also has to become more emotionally intelligent.

In the last couple of chapters I have reflected on some common aspects in our ancestors' descriptions of how invasion and annexation looked and felt in the deep past and I have highlighted some themes which have persisted into the modern era. Having traversed this darker territory, it is important to set this picture in its bigger context. Because, as we have already seen, in the world of ancestral story not every interstellar incursion is an invasion. When I survey the global family of paleocontact stories a complex and multi-faceted picture emerges. It is not a pattern of black and white. It is not that we humans are the goodies and the ETs are the baddies. Neither does the ET contingent divide neatly into a cast of friends and enemies.

In the spectrum of life in the cosmos, yes, say our ancestors, there is the prospect of visitors who may occupy the draconian end of the spectrum, just as Kam says of the Mo-o and Ahumanu. But at the lighter end of the spectrum there are nurturers and helpers whose positive impact on the human story has been incalculable. And from the Anunna of Sumeria to the gods of Greece, we hear of conquerors and saviours and every colour and shade in between those two poles. In ancient Armenia, for instance, our friend Haldi carried medicines, seeds and the pinecone of ascension. But he also carried his *suri* – his fearsome weapon. In ancient Sumeria, Enki saved and helped humanity, yet he was still the genetic master who toyed with our DNA and shortened our lives. Through the lens of the Popol Vuh, Kukulkan toxified our ancestors' environment, yet without him we would not be the beautiful, creative species we are.

Taken all together, ancient paleocontact narratives are a kaleidoscope of colours. If our world mythologies cast an accurate

211

light on exopolitics today, then we need to be ready for more subtle conversations and move beyond a *"friend or foe"* mentality with those with whom we are in contact, if we are to achieve any real sense of communication and discernment. That's where we are headed a few pages from now. Meanwhile, as I sit at my desk, staring out at the wildlife on the lush hillside, I wonder how many familiar political and social conventions we may have modelled on past contact, and how much might indicate that we have already been well and truly invaded. Thinking about the alleged treaty at the time of presidents Truman and Eisenhower I wonder how those agreements might be playing out in the present. If there really is a compact in place dating from that time, as Haim Eshed has claimed, how similar might those agreements be to the deeds signed by Queen Lili'uokani when she *"agreed"* with the USA, that the USA could annex her country? Are we still bound by exopolitical decisions made in the 1940s, or do we have more help at hand and more options available in the twenty-first century? But right now my eyelids are getting heavy and I really need to turn in. My not knowing will have to continue a little longer.

In the next few days, I am anticipating a phone call from a fellow researcher and author in Montana, a source close to President Eisenhower, who might have something to offer to inform my questions and speculations. These questions can be overwhelming and can make the ordinary human on the street feel disempowered and irrelevant and this is why in the next chapter I am going to take you with me to Europe and to a most beautiful mountain range for a refreshing skiing holiday. Here a surprising encounter with an ancient language will direct us to some intriguing tranches of human history.

Walk with me through these historical vistas and, I promise you, we will come away with reasons for hope, and indications that we, the human race, might have far more support, and far more resources up our sleeve with which to adapt and thrive than we might ever have imagined, whoever or whatever the one percent of the one percent might be. If this is of interest then I invite you to join me in the next chapter in the beautiful snowfields of the Italian Tirol.

CHAPTER FIFTEEN

Ancestors and Orchards

The Dolomites, Italy

Breathing in the clear mountain air of Val Gardena, I feel like I could live forever. I can't quite put my finger on it, but something here makes me feel profoundly at home. I wander the streets of Val Gardena deeply inhaling the pristine air, contemplating the white ski slopes, the verdant pastures and the white and blue-grey mountainscapes. I think I could live long and very simply in an environment like this and yet feel incredibly wealthy. The feeling catches me by surprise because really, I am not a mountain person. I'm a beach person. Yet there is something about this place which feels strangely like home.

It is 2003 and I am at the mid-point of a thirty-three-year journey in the world of ministry and I'm enjoying a long overdue break in this spectacular and elevating place. My period in church-based ministry has already taught me that senior positions in the church world can be incredibly rewarding professionally while at the same time tremendously unrewarding, financially. Outside of the world of church, I think most people would be taken aback by how many senior pastors are expected to self-fund and self-cannibalize in order to support the institution. Indeed, I think it is in a spirit of empathy for a brother enjoying a leaner chapter that my friend Jez has generously gifted me with this wonderful skiing holiday in the Italian Tirol. It is the most generous and thoughtful gift, exceedingly well-timed, and I am feeling rejuvenated just by being in this beautiful environment. It is the perfect place in which to simply breathe.

It has been a few years since I have made any attempt at cross-country skiing and today I have been pleasantly surprised to discover that my skiing skills appear to have lifted since my last tentative foray as a teenager in the Canadian province of Quebec. If you are wondering what exactly we are doing here, I have an ulterior motive for inviting you on this particular skiing trip. I am about to take you down a slope which on first inspection may seem quite removed from questions of ET invasions and for a page or two you may feel that you have slipped into an alternate reality and are somehow reading a book of imperial history, but stick with me and you will soon see why we have come this way. Follow me down this piste for half a chapter and you will see how the lessons from Kam's telling of the Mo'olele percolate through the *Pa'an* (geneaology) of all our families, shaping the fortunes and migrations of our ancestors, yours and mine.

The town of Ortisei, in the alpine valley of Val Gardena, has many attractive features as a tourist destination. It also has a claim to fame which pricks my ear as an addictive linguist. Over a toasted cheese sandwich lunch (the staple lunch of the region) with my local guide, Gianna, I mention in passing that somehow, I find I am able to decipher much of the local language when faced with the local street signs, menus and local literature. I am just not sure how I am doing it. For four years at school I labored at learning the German language. So being able to decipher the signs written in German is no surprise. Subsequently I learned Italian at the University of Bath and the Istituto Machiavelli in Florence. So the signs in Italian also pose no great challenge. However what I can't get my head around is why the third language in use here is also familiar to me. I just can't place it.

Gianna holds the ley to this little mystery. The story she is about to share with us speaks to many of the questions raised in the last couple of chapters:

How can you and I thrive if the decisions which shape our world are between a "secret commonwealth" and an unaccountable one percent, with you and me well and truly out of that loop?

How can you and I prosper when today's El-Ba'adat appears to treat the vast majority of us as tenants in someone else's economy?

What Gianna has to share will remind us that we are not the first generation to be asking questions like these. So, back to the mystery of the menus and street signs.

"Paul, did you study Latin at school?"

I have to confess to Gianna that as a teenager I endured three years of escalating stress, laboring to get that particular language under my belt.

"Well Paul, that is why you can read the local signs and books. The language we speak here is called Ladin, and it is directly descended from the Latin language which was spoken throughout the Roman Empire two thousand years ago. Ladin is Latin's closest living relative in the world today. And I will tell you why it is spoken here."

The amazing story that my Ladin-speaking guide now relays to me is astonishing and makes me feel incredibly connected to seismic geopolitical events one and a half millennia in the past. By the end of the chapter you will see how the history of this place half a world away from Hawaii relates to Kam's telling of the Mo'olele and how his ancestors' responded to the Mo'o takeover. Gianna's story takes us back to the fifth century of the common era, when Rome was actively withdrawing her material, military and administrative resources from all the western territories of the empire. One hundred years prior, Rome had identified Constantinople as its new capital. Constantinople would provide a centre for the empire which was more strategically sited for a global government, in a location which straddled the East and the West. Finally, the moment arrived when the division of Rome's army stationed in the Italian Tirol, based in what is now Ortisei, received its marching orders. They were to decamp and travel east to regroup in the new capital.

"When that moment came, the entire Tirolean legion decided en bloque that they would not comply. They all deserted. They all agreed to stay put. The empire had lost all interest in maintaining order in these territories and so the Tirolean division banked on simply being left alone."

"With the few local people of the area, the Roman soldiers and their families built a new life for themselves here. Instead of serving the global imperium they now looked to their own and each other's interests and put their energies into building a community worthy of their sweat and loyalty."

"And so, through combination farming, craft and hospitality they created a self-sufficient economy in the region which, as you can see, continues to thrive today. We have been here ever since. This is why you find us still speaking almost the same language that we did nearly fifteen hundred years ago, back when we made the decision to stay and make this place our home."

As I listen to Gianna's ancestral memory, spoken in her lilting Ladin-Italian accent, it isn't hard for me to sympathize with those Roman military families in the four hundreds, whose descendants live here to this day. If I had been among their number I can well imagine looking around at this fertile and beautiful country and thinking, *"Yes. There is enough here for us, working together, to create a better life than if we were confined to someone else's eastern city, slaving for an empire that has all but forgotten us."*

The experience of empire of the Ladin-speaking people in the Italian Tirol repeated in other places throughout Constantinople's western territories, as citizens increasingly found themselves to be still taxed but neither resourced nor represented – a scenario with an historic resonance for my American brothers and sisters. In the decades prior to Rome's military withdrawal from the West, the large structures of government had already dialed down their commitment to infrastructure, services and protection compared to that of previous generations. The era of empire-building and imperial nation-building had done its work in the west and was now on the decline. By this point, the interests of the Empire's ruling factions, the one percent of the Pax Romana, had become impossibly removed from the interests of its citizens in western Europe. Though the *"New Rome"* in Constantinople continued to enjoy a steady trickle up of western wealth through its matrix of trade deals, commodity pricing, taxation, and the control of the empire's single currency, it seemed that citizens on its western fringe were enjoying less and less of a stake holding in the imperial economy with every passing year. Does any of this sound familiar?

When Rome finally withdrew her military and administrative arsenal from western Europe, the citizens of those regions had no choice but to do the same as the Roman legion in the Italian Tirol with whose great-great-grand-daughter I have just enjoyed lunch. This geopolitical shift forced the people to find ways to recreate more local political and economic expressions. *"If Rome won't help us, then we will just have to help each other."* This became the order of the day.

It was within this social vacuum that a new kind of collaborative community began to appear in the mountain country of Italy. These communities followed a pattern of life developed by a young scholar living in the high country of Monte Cassino, eighty

219

miles southeast of Rome and eighty miles north-west of Naples. The young man in question is known to history as Benedict of Norcia. In modern terms Benedict was an academic drop-out. He rejected the conventions of sixth century Roman Catholic life and threw his energy instead into modelling an approach to life in community. Those who followed his approach, a number which mushroomed in the years following his death, provided the post-Roman West with a phenomenal network of social entrepreneurs, and a Europe-wide web of what were called *coenobitic* communities. (The word *coenobitic* derives from two Greek words meaning *"household life."*)

The coenobitic communities of the sixth century comprised monks, nuns, families, tenant farmers, scholars, researchers, scientists, engineers, medics, writers, students and employees. Working together, these communities generated strong local economies for the districts they served. Their economic enterprise centered on local farming and manufacturing. Yet these social pioneers did far more even than that. Their coenobitic villages became Europe's local nodes of international communication, literacy, education, agronomy, medical science, materials science, civil engineering and technology. Schools with education in them, farming with agricultural science in it, medicine with the most recent progress in it, hospitals with patients and nurses in them, libraries with books in them, pubs with beer in them, windows with glass in them: these all came to the de-Romanized territories of western Europe through the enterprise of communities following Benedict's model, which had codified a way to organize and create new local patterns of economy. It was social enterprise on a scale which was desperately needed in that period when the larger structures of New Rome had abandoned any commitment to its western citizens.

It is not difficult to find parallels with the kind of disfranchisement and disinvestment that people in many western countries are experiencing today, especially those who live any distance from the major cities. Once again, if I may, I will use my own country, Australia, as an illustration from the present day.

Taken as a whole, Australia is the 12th largest economy in the world. It is home to 0.3% of the world's population. Yet it holds 1.7% of the world's wealth. Its gross domestic product is $65,000 U.S. per capita ($100,000 AUD.) According to the investment bank *Credit Suisse*, the median Australian income in 2023 is higher than any of the G20 countries, at $273,900 U.S. ($390,870 AUD.) That is three time the median income in the U.S.A.

Yet 3.3 million Australians live below the poverty line - that's one in every eight Australians living in poverty. From Christmas to Christmas the only way these one in eight Australians are able to make it through the season is through the work of charities rattling collection tins outside supermarkets, and charity stores, known as *"op shops,"* which recycle clothing and toys so that poor Australians will at least have something. Major supermarkets place boxes and barrels outside their stores where more comfortably off Australians can deposit non-perishable foods to keep hungry Australians from starving. For this to be the lived experience of so many Australians, it would seem that those impressive median income figures must be somehow skewed by a few Australians who are fortunate enough to be very wealthy indeed.

In my previous state of Victoria, my church was part of a district-wide initiative in which a wide range of churches came together in the winter to staff overnight accommodation for Australians without homes. Under the leadership of the premier Daniel Andrews, the state government had recently illegalized sleeping on streets and empowered the police to confiscate whatever meagre possessions might still be owned by a person caught sleeping on the street, including any spare clothes and sleeping bags. In addition to the problem of street-sleepers, deprived of their possessions by the police, our state also faced the growing problem of families with two working parents, who were finding themselves priced out of home-ownership, priced out of rental accommodation, or disqualified from the rental market as a result of never having rented before and therefore lacking the necessary landlord references. If such working families were found to be living and sleeping in their cars then they were not counted among the statistics of the state's homeless, since they were, according to the government, accommodated.

In the absence of any other provision, the churches stepped forward and made their church centres available for those with nowhere else to go. The churches and social entrepreneurs involved worked hard to ensure that everything was done correctly and in compliance with all the byelaws protecting the safety of people and property. In addition to this initiative, local landowners offered to make their lands and properties available for emergency shelter, and temporary housing. It was a social collaboration that was truly heartwarming to see and for many it not only enabled them to survive the winter but revived their faith in humanity.

In their enthusiasm, the leaders of these initiatives proudly showed our local parliamentarians what this wonderful level of collaboration had made possible. Having been shown, our local government acted immediately and by the time the next winter came around they had changed the law, cancelling all the legal provisions which had made these collaborations possible, and illegalizing all these avenues of immediate, practical help.

As one of the pastors involved, I was disturbed by this response, so much so that my concerns led me to the office of our local member of parliament, a senior figure in the government of the day. I gently mapped out three simple amendments to recent legislation which would begin to correct the housing problem. To my astonishment, our member of parliament responded by saying with absolute clarity, *"Paul I don't see that government has any meaningful role in addressing housing needs. I see that really as an area for churches and charities."*

I politely pointed out that, unlike governments, churches and charities cannot legislate, nor can they classify real estate, alter zoning laws, regulate the work of developers, or release land and property for residential development. All these things are the preserve of government. I tactfully reminded him that the churches and charities had in fact just been legislated against specifically to prevent them from addressing the housing problem. Suddenly my noble friend remembered another meeting he had to get to concerning *"more pressing matters."*

At this point, you might be wondering why I am telling you all this. Simply because, given Australia's fabulous wealth, is any other illustration needed to show how the very idea of government has become totally removed from any idea of looking after the hard-working, tax-paying people whom the government supposedly represents? Just like our western European forbears in the New Roman Empire of the fifth century, we appear to be looking today at a world of taxation without representation, a world where the interests of government and commerce are increasingly divorced from the interests and needs of the grassroots population. As Gianna describes the pivotal moment when her Roman ancestors chose to stay and build a permanent home together, I think I can imagine what it must have been like as they looked at each other and said, *"Rome has clearly forgotten us, so let's work for each other and look after one another here. Together we can make it work."*

As I picture the moment, the twenty-first century café in Ortisei disappears. The combination of Gianna's story and my toasted cheese sandwich has triggered a Proustian cascade of memories, or what my family calls a *"Ratatouille moment."* If you're not familiar with the movie of that name, there is a scene in which a world-weary food critic is suddenly captured by a flavour and transported to a moment in his childhood, back to a place of love and security. I want to share my Ratatouille moment with you because it shows how Gianna's ancestral story of the Tirol resounds with some universal themes.

I am standing in the house of my maternal grandparents in the Chiltern Hills of South Buckinghamshire, England. This is the English, Welsh and Viking side of my *pa'an* and the sights, sounds and smells around me are those of a different world. My maternal grandmother, Reta, grew up in a home with packed earth floors, no electricity, and no plumbing. Every morning, each household would fetch its water from the village pump. Oil lamps and gas lamps provided the lighting. The cooking and heating were with solid fuel, and once a week the *"poo truck"* would come and empty the earth closet at the end of the long strip garden and cart it up to London for processing. No refrigeration meant a quite different pattern of eating and preserving food, and though money was certainly earned and spent, it was not needed for a good proportion of the household's weekly food intake, which was amply provided for by a garden full of fruits and vegetables. In fact each strip garden was filled with roots, vegetables, nuts, and fruits. One neighbour kept chickens, and another goats. Combine the gardens and it was like living on a small farm.

On the weekends and summer holidays when by brother Mark and I would stay at Gran and Gramp's I would find the mysterious accoutrements of this generations-old lifestyle, scattered around their semi-rural property. The ancient butter churner and the evaporator were like museum pieces for me, having grown up with all the benefits of electricity and supermarket shopping. My gran's pantry was stocked with fermented foods and pickles in sealed storage jars. Other jars were furnished with leaves, infusions and other mysterious contents for any immediate first aid and healthcare needs. This array created a unique aroma, once smelled never forgotten. Gran and Gramp's garden wasn't huge and yet it produced more food than we could ever eat. Their orchard and vegetable patch served our needs, and the surplus, which would only spoil if it wasn't used, was simply shared with the neighbours. A young neighbour might call in on passing, and Gran would say, *"I've got some apples and potatoes here. Take them home for your mum."* Another neighbour might pop her head round the back door. *"Paul, see if your gran can use these eggs."*

It was not only food that was shared. Gran and Gramp's house had two small rooms downstairs for living and three bedrooms upstairs. Yet, back in the 1950's my mum's cousin and her parents occupied the front living room downstairs and the front bedroom upstairs. Gran and Gramp had the second bedroom and Mum would often share the third bedroom with one of the girls from next door, whose family had seven members to accommodate in an identical amount of space. This was nothing unusual. It was simply the sharing of life. The little row of identical and mirror-image duplex homes had been inhabited largely by the same families since they were built in the early twentieth century, and so by the time I came along and was introduced to the picture, this little community of neighbours had navigated some very tough years together, the years of the depression in the late 1920s and 1930s, the years of war in the 1940s, and the maintenance of wartime privations in the form of rationing until the mid-1950s. It was only decades later that I was able to appreciate that it was this informal shared economy which had enabled these households to navigate such difficult years, and together survive and thrive. As

Gianna describes the founding of Ladin-speaking Ortisei, I realize I have seen a glimpse of that same collaborative spirit in my own growing up. It's not that 1920's-1950's Great Britain was an idyll. Far from it. They were tough times. Yet there was clearly something other than the spirit of competition and self-reliance operating in the culture of that street in those decades.

In the light of all that, I wonder if finding each other in this kind of way, and rediscovering more local expressions of economy might become important again in the decades ahead. If the concerns of our uber-governments become too remote, then at the level of household, friendship and family circles, neighborhoods and towns, we might need to recover the kinds of synergies and local economies enjoyed in previous times. These synergies are not the monopoly of sixth century coenobitic communities, or the post-war years or the years of depression of the twentieth century. Every generation finds its own ways and means.

In my own lifetime I have known the pleasure of sharing my home as an extended community, opening my home to a sequence of forty boarders in the years before Ruth and I were married. In the U.K. through the '80s and '90's I knew families who opened their homes to refugees and orphans when the powers of government had nothing to offer beyond *"solutions"* which would be only damaging to their recipients. I have seen one friend pay another's mortgage for twelve months when the national government had no solution to offer a depleted jobs market, escalating interest rates and people trapped by negative equity due to a burst bubble in the property market.

In the U.S.A. I have friends who have set up stakeholder companies in which the company is owned by its employees and the profits shared, in the opposite spirit of the prevailing culture of laissez faire, fat cat CEO's and shopfloor workers on benefits. When larger scales of economy have failed our farmers, friends in small towns have banded together to make themselves the new corporate customer buying directly from the farms in order to keep their farmers going and their local communities thriving.

These are not revolutionary actions. Nor are they rocket science. We all know how to create and occupy the kind of local economies which can step in where larger structures have failed to meet a need. Whether we look at the level of household, family, neighborhood or town there is always scope for empathy, imagination, compassion, solidarity and creative enterprise to chart a way forward. I share these examples as personal reminders that whatever the One Percent are about, we at the grassroots can make choices which generate movement in positive directions. When the larger structures of government under-invest in the human-scale economies of our regions, that is precisely when you and I need to move in the opposite spirit and invest ourselves into local and peer-to-peer synergies, grassroots collaborations and local economies. If the higher-ups appear to forget us, that is when we need to rediscover one another. So much of my own work in the last few years has grown up in this kind of new economy.

Everything I have just told you, my stories of fifth century Ortisei, twentieth century South Buckinghamshire, and the U.S.A. in the twenty-first century, I now relay to Kam in his lush, green conversation pit, and when I do, he absolutely lights up.

"Paul, what you have just described, that is how it was before the Mo'o came, before the arrival of the Ahumanu. It was a world where we were not divided against ourselves. We were a community."

As we talk, Kam echoes back to me all the themes I have elaborated on in this chapter. Every invasion and annexation in his people's history has involved the hijacking of the larger structures of economy, politics and law, forcing his ancestors to rediscover

227

each other and rebuild their culture at the grassroots. Collaboration and community are in Kam's blood and the key to that, he says, is not some clever political strategy or a legislative hack. The secret, he says, is a deep nurturing of our empathy towards one another as human beings.

"We were a community not because of a strategy or a code like Benedict's, but because our manawa was open. This meant that we knew each other's thoughts, and we felt each other's concerns. When people truly see each other, they cannot live against one another. When the manawa is open a person has no choice. We had to find ways of harmony."

"Before the Mo'o came, the land had not been divided and privatized. So homelessness was impossible and the land was fertile. So this meant nobody ever needed to go hungry. And that is how we lived, helping and serving one another, and in this way the whole community thrived."

As the light begins to fade and the evening falls, I do my best to relay to Kam all the Italian, British and American stories I have shared with you in the last few pages. I want to tell him about my pa'an, my grandparents on my English, Welsh and Viking side, and the other things I have seen personally, because I want to tell him that I know that the harmony and community he describes in the Mo'olele is not some impossible idyll. I assure him that I understand that the peaceful existence he describes was not some magical gift that had supernaturally dropped into his ancestors' laps. I appreciate that their peace must have been hard fought for and hard won, and that the equilibrium his ancestors finally achieved would have been all the more valuable to them when seen against conflicts in the times before and after.

I want to reassure my friend that I know from my own pa'an on my European and African sides, that empathy and solidarity can produce solid, practical results, even in the most challenging of times and so I do not listen to the Mo'olele as if it were a narrative of pure fiction. I am pleased to see Kam smiling and clearly clicking with my stories of other times and places and I soon have him on a roll too, happy to be sharing his wisdom with an eager student.

"Paul, for you and me to thrive in a society that has been hijacked by inhuman forces, the question is really about how human you and I want to be."

"What does it mean, to be human? It means to live intelligently and from the heart. So, seek a heart connection with your friends, your neighbours and your local community, and find ways to help each other at those levels."

"Whatever skills you have, make sure you pass them on to the next generation and to anyone who needs them. Grow your own food. You only need a small strip of land to feed a family, and if you grow fruits and veggies, you will always have more than enough to share with others."

"Your environment has everything you need to support your body and mind. So make sure you are living in the environment, not locked away indoors, but outside in nature every day. If you are in nature less than half an hour a day, that is not good. Connect with the sand and the soil, the plants, the sunshine and the fresh air. This is your basic medicine."

"Tell stories to each other which will help us all to remember that at heart we are empathic and compassionate people. Nurture your children's compassion through connecting with plants and animals and through all that you teach and model in yourself. Nurture their imagination and their intuitive knowing because

they are the ones who will have to adapt to conditions in the future which we can't predict for them."

"Nurture your own na'au too, and learn to keep your manawa open to intuitive knowing and help your young children to do the same. If we can live with an open manawa then we cannot be lied to, we cannot be deceived and manipulated. An open manawa means we can see things that are coming down the line so that we know how to adapt."

"Whatever the higher ups are doing, we can always find ways to serve one another and create our own environments."

Kam pauses.

"And you should move out of the big cities!"

This last suggestion is very topical for me. In fact, my family and I have done exactly that, moving first from the Australian Capital Territory to an outlying town, and then onto a quieter, calmer, regional town in sub-tropical Queensland. So Kam and I are on the same page. We are pleased to be closer to family, to have more time for friends and neighbours and more space to grow food. Our move is part of a significant demographic shift in Australia, one which has disrupted the national uber-policy I mentioned before of concentrating populations within a tight perimeter of the major cities in ever-denser housing, ill-suited to activity, outdoor life, contact with animals and nature, and deleterious in general to mental health. The government's previous intention to deprioritize schools and hospitals in our quieter, more liveable part of the world has had to be reversed as tens of thousands have moved into the region, creating demand for new infrastructure. It is refreshing to see our government having to respond to the grassroots and it reminds me that not everything is set in stone.

As the lands of Hawaii disappear beneath the clouds and another flight carries me to cooler climes, I feel pensive about what I have seen and heard in my time with Kam. What has really struck me is that when he speaks as a guardian of ancient stories of paleocontact, his concerns are framed very differently to today's chatter in the world of Ufology and Congressional hearings. For Kam worrying about ET invasion is not about watching the skies for anomalous craft or scanning YouTube for videos of unexplained sightings. Instead, he is concerned about shifts and changes in our way of life. His questions are about the directions in which our culture is moving. Are we moving in a human direction? Are we moving towards compassion or machinisation? Are we moving towards harmony or are we being divided and ruled? Is human imagination being fostered or is it being shut down? Are we getting dimmer or smarter, healthier or sicker?

Kam's vision for moving things in a human direction is of a shift from division and competition to empathy and harmony; from selfishness and greed to patterns of community; from slaving for masters to working for one another to; from laboring for money, to enjoying the fruits of the earth. And I note that he isn't sitting on his farm, twiddling his thumbs, looking to the politicians and higher ups to be the gatekeepers of our future. He is very focused on what you and I can do, what emotional intelligence you and I can bring to the party.

Kam's words to me have reminded me of something President Eisenhower said back in the day. Something about protecting a *"way of life."* It makes me reflect that perhaps knowing who or what exactly could be influencing the One Percent, or who or what is oversighting The Program at the Pentagon, is ultimately less important than what kind of culture you and I are building in the places where we live.

I think Kam is right. In every era of human history, people have had to find ways to thrive in a changing and unpredictable world, and in every age we have always been able to adapt. We really are *Homo Sapiens Adaptabilis*. What has impacted my personal confidence more than anything as I look to an uncertain future is the chorus of voices from indigenous and shamanic traditions, like Kam's, which tell me that humanity has made none of this journey alone. We have always had company and we have always had help. It is a lesson for me personally which is about to repeat in another mountainous region thousands of miles away, in the foothills of the rocky mountains of Colorado.

At a staging post en route, I find myself sitting with anticipation in front of my laptop screen. I have a conversation scheduled with a fellow author in Montana, who I hope might have some privileged information to cast light on that seventy-year old story about President Eisenhower and his alleged expolitical inolvements. I really want to know if that story is true. Haim Eshed's statement, naming that more-than-seventy-year timeframe would certainly appear to authenticate it. If it is true, where did the knowledge of ET compacts take the president in the years that followed? What does a person do next when their entire universe has just changed? My laptop is trilling. It's time for that call.

CHAPTER SIXTEEN

Calls and Contact

Montana, U.S.A.

"How could anybody process the larger picture of all this?"

I have Laura Eisenhower on the line.

"My great-grandfather and his trusted circle were well-intentioned people trying to do their best for humanity, aware of what was going on to a certain degree, but heavily stopped from disclosing."

Laura is the daughter of Susan Eisenhower, who is the granddaughter of President Dwight *"Ike"* Eisenhower. Eisenhower's alleged contact with ET visitors in 1954 is not a topic the president's descendants like to discuss. Perhaps the matter is too overwhelming. Maybe they don't believe it. Laura is the exception. She has been willing to lift the lid on the 1954 story and attempt to disentangle fact from urban legend. Conclusions Laura has drawn from her own sources match my own analysis of that story, namely that the shape of our current exopolitical situation was established prior to Dwight Eisenhower's election to the presidency, having been defined by covert military intelligence provided to President Truman. To Laura, the truth of this matters at a profound and personal level, as she identifies deeply with her famous great-grandfather.

Before he became America's thirty-fourth President, Laura's great-grandfather commanded the allied forces in Europe during the Second World War. By all accounts he was a man of tremendous courage and unimpeachable integrity. How would he have lived, I wonder, with such a great burden of secrecy upon him? If he had privileged knowledge of cosmic visitors, collaborators or overlords, where did that knowledge direct his attention? And did the President find a way to speak his mind without breaching the layers of official secrets laws which would have threatened him and his family for generations should he ever trespass into revealing forbidden knowledge? In answer, Laura points me in three directions.

"Firstly, my great-grandfather felt free to speak openly about his own 'personal belief' that we are not alone and that we are raised by 'greater forces.'"

In this kind of language I recognize the same formula which Apollo astronaut Ed Mitchell was to use decades later, as he called on the powers of government and military intelligence to declassify what they know about our cosmic company, without going quite as far as stating forbidden facts. Whenever Ed Mitchell directed his attention toward those who are skeptical of UFOs and contact phenomena, he would always highlight one particular white rabbit, which he was confident would lead anyone who followed it to the conclusion that we are well and truly in contact. That white rabbit was the scientific examination of UFO retrievals. Before his death in 2016, Ed Mitchell considered there was already enough documentary evidence in the public domain surrounding recovered materials to convince any sincere skeptic that extraterrestrial visitors and their craft are objectively real and worthy of serious study. Within three years of his death the Pentagon's public acknowledgment of The Program has finally validated Ed Mitchell's assertion.

The second white rabbit Laura now points me to is couched in the text of Eisenhower's ten-minute farewell address to the American people, televised on January 17th 1961. In this famous and much-studied broadcast, the president's concern was not to do with any threat of ET invasion but rather the grave danger of over-empowering the military and the business corporations who profit from the technologies of war. This is what he said:

"In the councils of government, we must guard against the acquisition of unwarranted influence, whether sought or unsought, by the military industrial complex. The potential for the disastrous rise of misplaced power exists and will persist."

"We must never let the weight of this combination endanger our liberties or democratic processes."

"We should take nothing for granted. Only an alert and knowledgeable citizenry can compel the proper meshing of the huge industrial and military machinery of defense with our peaceful methods and goals, so that security and liberty may prosper together."

With these words the president fixed a spotlight on the very same corporations we identified in chapter twelve, whose special access programs would be deeply informative in our quest for disclosure, if only they had the liberty and the incentive to speak. Notwithstanding Deputy Secretary of Defense, Kathleen Hicks' verbal commitment to *"transparency"* and *"sharing...with Congress and with the public,"* it would seem that the embargo over these operations has not been lifted. If it had, I have no doubt that my compatriot Ross Coulthart would have been to the labs concerned like a shot.

When Eisenhower pointed his finger at the military industrial slash aerospace community, it was not The Program as such nor the prospect of ETs in the Whitehouse that concerned him. Rather it was a shift of culture which he feared the military industrial community was already effecting on our politics, domestic and international. His concern was that this powerful community of corporations, if unfettered, could foster a culture of perpetual paranoia and war, as dramatized by George Orwell, from which it could then profit on an even greater scale. Eisenhower's direction was that we should be be quick to recognize and guard against any kind of dilution of our democratic freedoms to a world of centralized power in the name of *"defense"* or *"national security."* This was a surprisingly pointed statement, all but naming the corporations and, by implication, the families who own them, whose influence over our geopolitics was, he said, *"unwarranted,"* *"misplaced"* and potentially *"disastrous."* Eisenhower didn't name names because, after all, names can change from one generation to the next. Rather he highlighted as clearly as possible the persistent danger of corporate powers whose economic incentives run counter to our *"liberties and democratic processes."*

The third white rabbit that Laura highlights for me to follow is to be found burrowed in the president's parting words about our need at the grassroots to be *"an alert and knowledgeable citizenry."* What did Laura's great-grandfather mean by these carefully chosen words? Alert to what? Knowledgeable about what?

"I think that was my great-grand-father's way of telling people not to accept incomplete or shallow explanations and not to overlook misdirection from any source - including from government sources. He was telling the people not to be credulous, but rather to be curious and always to dig deeper."

There's no missing it. Eisenhower's words amount to a warning to fully expect smokescreen governance and official misdirection. It is *"the citizenry"* whose knowledgeable and alertness will *"compel"* accountability and course corrections.

As it happens, the farewell address of 1961 was not the first time President Eisenhower had identified a worrying tension between goals oriented towards human wellbeing and the quite different

236

goals of the military industrial complex. In 1953 and 1954 Eisenhower made these solemn statements:

"Every gun that is made, every warship launched, every rocket fired signifies, in the final sense, <u>a theft</u> from those who hunger and are not fed, [from] those who are cold and are not clothed. This world in arms is not spending money alone. It is spending the sweat of its labourers, the genius of its scientists, the hopes of its children. The cost of one modern heavy bomber is this: a modern brick school in more than thirty cities. It is two electric power plants, each serving a town of sixty-thousand.... It is two fine, fully equipped hospitals. It is some fifty miles of concrete highway."

"<u>We pay</u> for a single fighter plane with a half million bushels of wheat. <u>We pay</u> for a single destroyer with new homes that could have housed more than 8,000 people. This, I repeat, is the way of life to be found <u>on the road the world has been taking</u>. <u>This is not a way of life</u> at all, in any true sense..."

In another setting, Eisenhower stated what he believed ought to be the proper goal of the military industrial complex along with every other human community and institution. He said, *"We do not keep security establishments merely to defend property or territory or rights abroad or at sea. We keep the security forces to defend <u>a way of life</u>."*

(My underlinings.)

These were not the words of a coward or a faintheart. They are the words of a five-star general, the commander of the allied forces in the Second World War. What concerned President Eisenhower the most were militarizing forces which would seek to centralize power and wealth and move society away from *"a way of life"* in which government is centred on feeding the hungry, clothing the poor, honouring the labour of its people, the genius of its scientists

237

and the hope of its children. His aspiration was for government which would invest its great resources into providing power to its populations, fully-equipped hospitals for its people and schools for its children. For Eisenhower, the gift of government meant the power to distribute highways to connect the country's people, plentiful harvests to feed its people, and homes to house them.

In other words it would seem that Eisenhower, in his later years, was less concerned with any supposed military threat from international or interstellar neighbours, and more with defending our way of life against cultural shifts, which were already resulting from the leveraging of *"national security"* fears and *"existential threats,"* whether real or spurious. This is a perspective which might serve us well in the present hour as the language of national security and existential threat enters public conversation once more in all the furore around the Pentagon's embargoed information about UFOs and UAPs.

I return to Eisenhower's hope in *"an alert and knowledgeable citizenry,"* and I ask Laura if she feels the same way.

"I do. That's why I do everything I can think of to encourage people to be curious and to dig deep," she says, *"To resist the dumbing down of society and pursue personal alchemy and ascension. I want to help every one of us to remember who we are as human beings, to recognize the greater forces that raise us and that want to lead us to live in harmony with one another and with the planet."*

To her credit, this is exactly what Laura Eisenhower is doing with her life's work. Indeed it was on a panel at a conference hosted by Neil Gaur's *Portal to Ascension* that I first encountered Laura just a couple of years ago. Her work centres on nurturing people's intuitive and empathic skills, fostering all the higher cognitive powers which my friend Kam would associate with an open

238

manawa. These are the skills of far sight and future sight, empathic and telepathic connection, inner sight and self-healing, about which our ancestors spoke and around which our ancient shamanic ceremonies revolve. (You can read more about these in *Echoes of Eden*.)

As I reflect on my conversation with Laura, it strikes me that Laura is not the first person to make a journey from being confronted with the reality of cosmic contact to the pursuit of higher cognition and heightened sensitivity. Plato made exactly the same journey Judging by his disclosures in *Phaedo* and *Timaeus and Critias*, Plato's exposure to contact phenomena, assisted by the Eleusinian, Orphean *or* Hecatic sects of his city, was a paradigm-shifting experience which informed his lifelong commitment to raising the consciousness and intelligence of all those who would come under the influence of his teaching in the ages to follow.

This transition from personal epiphany to the pursuit of heightened consciousness is the same pathway trodden by the late, great Ed Mitchell. Though he never spoke of contact phenomena on his own account, he did speak about the personal impact of a pivotal experience in space. This is how he described it:

"I had completed my major task for going to the moon, and was on the way home, and was observing the heavens and the earth from this distance, observing the passing of the heavens. As we were rotating, I saw the earth, the sun and the moon and a three-hundred-and-sixty-degree panorama of the heavens."

"The magnificence of all of this, what this triggered in my vision, in the ancient Sanskrit is called 'samadhi.' It means that you see things with your senses the way they are, but you experience them viscerally and internally as a unity, a oneness."

"All matter in our universe is created in star systems. And so the matter in my body and the matter in the spacecraft and the matter in my partners' bodies was the product of stars. We are stardust and we are all one in that sense."

"I realized that the story of ourselves as told by science—our cosmology, our religion—was incomplete and likely flawed. I recognized that the Newtonian idea of separate, independent, discrete things in the universe wasn't a fully accurate description. What was needed was a new story of who we are and what we are capable of becoming."

This experience of Samadhi drove Ed Mitchell's life's work for the next half century. His great obsession was in learning how to raise our human abilities in intuitive knowledge, what the Hermeticists called *Nous*, and what Kam calls *Na'au*. To speak of far sight or remote viewing, future-sight or precognition, heightened empathy or telepathy, telekinesis or self-healing, may all sound like so much fantasy, new age-ism or old-world superstition. Ed Mitchell called it *Noetic Science*, the science of deep consciousness, and to this day it is the goal of his *Academy for Noetic Science* to bring the objective rigour of modern scientific enquiry to the very deepest experiences of human consciousness.

My own journey of research in ancestral traditions has led me to believe, just like the late Dr. Mitchell, that our place in the cosmos relates directly to a true understanding of our origins as a species. For me, the personal discovery of a different story of human origins, as embedded in the world's ancestral narratives, has convinced me of a different view of human life and human potential. It's noteworthy that the same indigenous cultures which have curated stories of paleocontact have also nurtured mystical traditions and shamanic protocols aimed at accessing the higher cognitive powers inherited from our ancestors. The purpose of these protocols is for the shaman or guide to access their own cognitive powers and help their student or patient to unlock theirs and begin engaging directly with a higher field of information. This is Plato's *Sophia* or *Wisdom,* the Hermeticists' *Nous,* Kam's *Na'au,* and Ed Mitchell's *Noetic Science.*

The people I have listed in this chapter, whether President Eisenhower, his great-granddaughter Laura, Plato, Ed Mitchell, Kam, or myself, are not about questing for superpowers for their own sake. The purpose of *Na'au, Mana'o,* and an *open manawa* is a more intelligent and harmonious society. People who feel each other's joys and sorrows, who know each other's thoughts and feelings, who can foresee consequences for themselves and for others, are compelled to build ways of life that are embracive, harmonious, and sustainable. This kind of cultural recentering is desperately needed in the polarized and pugilistic world of today's politics. But before we start feeling too apocalyptic, we should pause and remember that neither these challenges nor their cures are new. As we will see in the pages ahead, we have been here before.

In the meantime, while you and I are busy working on our manawa, what is Robert Kirk's *"secret commonwealth"* up to behind closed doors? What *"research"* is Haim Eshed's *"galactic council"* into? What is up for discussion among the powers on the El-Ba'adat? Why exactly are they here? I don't have privileged information via military intelligence, neither would I claim to have far-sighted the answer in the way the *haro'eh* Micaiah did back in the day. What I can say is that our ancestors answered these questions long before you and I thought of them. Out of their library of narratives, in the next few pages I am going to put forward seven answers, not as proofs of anything, but rather as an overview of longstanding perspective concerning contact in our global canon of indigenous story. I will begin where my own paleocontact journey began, in the pages of the Bible.

Strategic Advantage and Access to Resources

One of the anomalies in the book of Genesis which first arrested my attention was the text's interest in locating the freshly generated human race close to key mineral deposits – gold in particular. Genesis portrays the first humans as living in a safe enclosed area, provided with every kind of food, medication, and comfort they could wish for. What, then, would such primitive humans need with gold? The additional note that the elohim who engineered the humans then set them *"to work"* raises the question of what work that could possibly be. Did that work have anything to do with the key mineral deposits I wonder?!

In the southern cone of Africa, mining companies have uncovered the structures of ancient gold mines dating back at least two hundred thousand years. Palaeontology says that our ancestors were here at that time, looking rather like you and me, yet not quite smart enough at that time to create a farm or build a city. Perhaps we were just smart enough to work in someone else's mine. Today we are just beginning to understand the technological and health-giving properties of gold, so our ancient visitors' interest in the planet's gold deposits perhaps need not surprise us. It is not necessarily that gold was the primary reason for ancient incursions, just as gold and cacao were not the primary reason for the British annexation of Ghana, and sugar was not the primary reason for the British annexation of the American colonies, or the Portuguese conquest of Brazil. The primary motivations were to do with control of the geopolitical environment, with the confiscation of copious natural resources as a nice material bonus.

Similarly, the Bible's sources, the Sumerian narratives of Enlil and Enki, suggest that the primary motivation of our visitors' annexation of Earth was dominance of this region of the galaxy. Terrestrial resources were a secondary concern to Enlil, the senior Anunna on the council. It is conceivable that Enlil's indifferent attitude to the human presence on Earth might mirror the view of some of our visitors in the present.

Exploitation of Genetic Material

The abduction and hybridization narratives of Genesis 6, which are elaborated further in the Book of Enoch, and which echo in cultural memory around the world all tell us that there is something about human beings which is uniquely attractive to our extraterrestrial visitors. Genesis 6 says it of the *benei elohim*, the Book of Enoch says it of the *Watchers,* and the Sumerian Enuma Elish says it of the *Igigi,* the *Observers.* These beings, observing the planet from space had, it seems, been doing some talent-spotting from afar and found human beings, females in particular, very beautiful.

Whether we listen to these narratives or the Greek, Norse or Indian stories which echo their themes, this particular wave of hybridization was remembered as a controversial breach of exopolitical accords already in place, in which our community of planetary observers had agreed on a guiding ethic of non-interference. The extraterrestrial forces who had already planted their flags on the Earth's surface had hegemony there and any other incursion was considered an unforgivable breach. It is worthy of notice that this element resounds in cultural memories of this wave of abductions all around the world.

The folkloric traditions of Wales, Scotland, the Philippines, Kenya, Ghana and the Caribbean, which we touched on before, all affirm that our non-human neighbours liked what they saw in the human race and wished to enrich their own gene pool with the features of human beings. From the perspective of our hybridizers it is interesting to speculate what it was they coveted and saw as our species' best features?

For instance, when I hear the story of the Mimi spirits described by Yolngu people of Australia's Northern Territory, they speak about the improbably slender physicality of the Mimi beings. It makes me wonder if our resilient and robust physicality might attract the attention of some of our neighbours with frailer frames. When I listen to the Celtic stories from out of Ireland or Scotland as they describe the behaviour of the *Sith*, the *"little people," "the other crowd"* or the *"good people,"* what I hear suggests beings who are similar to us but who lack the traits of compassion and imagination which make human beings special. All in all, it isn't hard for me to imagine that in *homo sapiens sapiens* there may be a unique blend of animal strength, mammal emotion, and higher intelligence which has made for a species with a unique capacity for creativity, intelligence, compassion and love. I wonder if our presence in the Milky Way has garnered the active interest of interstellar neighbours and cosmic cousins precisely because of the beauty and potential of that mix.

Another possibility, and potentially it's a more concerning one, is that other hybridizers may be here with a goal of making their descendants more human and consequently better adapted to life on Earth. Perhaps, in that sense, *Invasion of the Body-Snatchers* wasn't so far off the mark after all!

The Nurture of Humanity

Plato speaks in passing about the children of the gods. Their role in the story of humanity was, he understood, as upgraders of our ancestors' capacity for consciousness, curiosity and intelligence, effecting changes which would lead to a creative and technologically capable species. The modern theory of panspermia echoes Plato's thought that life is the rule in the cosmos, rather than the exception. To put it in contemporary terms we could say that the genetic coding for biological intelligence is as essential to the fabric of the cosmos as are the properties of light or gravity. This viewpoint makes all life in the cosmos ultimately related. The vital coding, wherever it finds a hospitable environment, a planet with water, can be expected to generate forms of life similar to those with which we are familiar on planet Earth.

By this understanding we would expect to find a universe-wide spectrum of older and newer cultures. Plato's notion of cosmic origins echoes in indigenous story around the world, with older cultures visiting our planet to nudge things along in the right direction, intervening after planetary cataclysms and supporting the development of human intelligence.

As for the progress of human intelligence, paleo-anthropologists in the twenty-first century are beginning to examine the possibility that our mental development as a species may have been accelerated through exposure to psycho-effective foods and drinks. Substances with hallucinogenic properties may have irrevocably altered our ability to access and play with our imaginative faculties and may have epigenetically redefined the cognitive abilities we would pass on to our descendants. Ancestral story and shamanic ceremonies around the world suggest that a mind-altering dietary shift of this kind was indeed part of our story as a species and that was a change made with external assistance.

In the Jewish and Christian scriptures these themes of external assistance can be found embedded in the motifs of Eden and the fruit of the tree of knowledge. It echoes in the Hebrew memories of Asherah and Dagon, in the Roman stories of Venus, the Greek stories of Aphrodite and the Egyptian stories of Hathor. In Zulu story it was Mbab Mwana Waresa and in Mesoamerica it was Hun

Hunahpu. These were the names of the ancient ones who came and taught our ancestors about the properties of various plants for foods, medicines and maybe for higher consciousness too.

In the world's oldest written narrative, The Epic of Gilgamesh, an advanced non-human tutor by the name of Shamhat nurtures the cognitive upgrade of the primitive human Enkidu, which she achieves through introducing the wild human to sophisticated foods and fermented drinks. Closer to the modern era in the world of ancient Athens, Plato's personal pursuit of higher consciousness led him to ingest the fermented tea kykeon, a brew similar to ayahuasca, with its distinct psychedelic properties. The context of Plato's experimentation with kykeon was a ceremony carefully curated with the intention of brokering contact with ancient cosmic tutors.

In a similar vein, the Dogon people of Mali, West Africa, speak of their own ancient helpers, whom they call the *Nommos, visitors* from the Sirius star system. The Cherokee people speak of ancient tutors who came from the Pleiades. On Knossos I learned about the role of the hybrid King Minos who elevated the culture of the Minoan people. In Australia my friend Djalu introduced me to the Yolngu people of Australia's Northern Territory, who speak of their ancient helpers, the Mimi and, as we learned on the shores of the red Sea, the Babylonians have their story of Oannes and the Apkallu. Every single one of these traditions regarded the news of a populated cosmos as good news, on account of powerful presences whom they had experienced as helpers of human progress, who by one means and made us more than we were before.

The Ascension of Humanity

Shamanic and esoteric traditions around the world take a still deeper look at who and what human beings really are. Plato captures some of this thought in his reasoning that before any of us become individual material beings we are each a fractal of that consciousness which is the source and property of the Cosmos itself. That is where our consciousness comes from. He argues that our bodies host and process consciousness but do not create it. Our participation in consciousness, he argues, is a participation in something that is endemic to the cosmos.

In a sense it is only a statement of logic to say that our individual consciousness is a participation in source consciousness and an aspect of the cosmos. Plato simply follows that thought to its logical conclusion.

By Plato's reasoning, this means that you and I as aspects of cosmic consciousness have come from a place of all-knowing into the discrete and finite experiences of the material realm, where in essence we have to discover everything from scratch. This epic of discovery, he argues, is therefore really a remembering, a recovery of knowledge which was previously ours, but which now as material beings we can learn to retrieve with the help of that aspect of ourselves which is most directly in touch with the cosmic source, and that is something beyond the conscious mind. We remember this every time our subconscious helps us to make an intuitive leap of some kind.

This may sound like a very rarefied and speculative line of argument. Yet this notion, carefully argued by Plato in his dialogues *Phaedo* and *Timaeus and Critias*, lies at the heart of Buddhism, Hermeticism and Catharism and echoes in shamanic and ascension traditions all around the world. When Kam speaks to me about Na'au and Mana'o, this is what he is talking about, perpetual cognitive expansion.

Through the ages this journey of ascension has been symbolised, modelled, and facilitated by great luminaries through the ages like

Jesus, Hermes, Buddha, and the Yellow Emperor. We might think of others who have advanced the history of human thought – Pythagoras, Ficino, Da Vinci, Giordano Bruno. What was it, I wonder, that gave these geniuses their edge?

Consider these iconic figures for a moment: Adam, Eve, Isaac, John the Baptist, and Jesus in the Jewish and Christian stories; Krishna in the Hindu tradition, Buddha of the Buddhist tradition, Laozi of Taoism, Zoroaster of Zoroastrianism, The Yellow Emperor, Pythagoras and Leonardo da Vinci. One feature they all share in their respective life stories is the element of an anomalous birth.

All are the product of artificial processes at the hands of non-human beings. With the exception of the archetypes, Adam and Eve, the stories of all their mothers include encounters with advanced beings and becoming pregnant by *artificial* means. Some are visited by a phenomenon of light, sometimes emanating from a craft in the sky, which then affects the body of the mother or the body of the foetus in utero in some mysterious way, resulting in the birth of a child with heightened intelligence, destined to make a critical contribution to the intelligence and consciousness of human culture. These anomalous conceptions and births produce people whose important contributions improve human wellbeing and provide new thought to enhance the social intelligence of humanity.

Now, let's jst take stock for a moment and survey the canvas of human progress that stretches from the moment we as a species could first think to the periods associated with contributions to higher thought, such as by Leonardo da Vinci, for instance. Along that canvas let me mark up all the contributions to human progress for which our ancestors gave the credit to advanced beings from the stars:

- Food Science

- Textiles

- Agricultural Science

- Fire

- Materials Science

- Land Science

- Writing

- Mathematics

- Civil Engineering

- Civic Administration

- Cosmology

- Higher Ways of Being

According to the texts and oral traditions of ancestors on every continent, every one of these advances was gifted to humanity by benevolent non-human visitors. This is what contact with advanced non-human civilizations was associated with in our ancestral past: progress in our way of life. What a different tone that is.

If the purpose of ancestral story is to provide us with a lens through which to see the present more clearly, then we have every reason to look at today's evidence that we have advanced company and in that realization find cause for hope and anticipation of a better future. This is not to be naïve or credulous. It is a matter of hearing our ancestors out when they tell us that there has always been external support for human progress to be sought out and found.

Support of Human Beings

Global stories of anomalous helpers, what we call *angels,* could also be said to fit in a scheme of advanced support. In fact, the word *angel* is probably an unhelpful or misleading handle, because though the word carries all kinds of spiritual and religious associations, at root the word does not convey any information about the biology, genus, or origin of the anomalous visitor. For example, the Hebrew and Greek vocabulary of the Biblical narratives simply indicates beings on a mission or with a message, who exhibit advanced capabilities which allow them to arrive and depart rapidly by means not understood by the contactee.

Origins stories such as those of the Dogon people, Cherokee, Mohawk, Mayan, Nigerian, Sumerian, Biblical and Aboriginal Australian narratives, all evoke a parental dynamic towards humanity from these visitors. For the Dogon, the Nommos of Sirius C are our ancestors from antiquity. They love human beings as their cosmic children. Perhaps this is our helpers' sense of Samadhi, as described earlier by Ed Mitchell, or we could understand it more concretely through the lens of interventions, genetic engineering and nurture. Anecdotally I find it curious that traditional descriptions of angels carry many points in common with ancestral descriptions of visitors from the Pleiades (more of which in *The Scars of Eden.*)

My travels through the East, in West and Southern Africa, central and South America have shown me that traditional cultures which have maintained shamanic protocols of healing and initiation build many of their practices on the expectation that our cosmic helpers are so warmly disposed to us as a species, and so intimately involved in project humanity that they can be contacted and tapped for information to support the health, wellbeing and ascension of individual human beings. Shamanic narratives around the world propose that communication from non-human helpers is not the monopoly of Admiral Wilson's *"corporate types"* but rather has been a mystical aspect of the human experience from the grassroots up since time immemorial, from the time of Haldi to the present day. It may surprise you to know this was also the belief of the very first Christians. This is a theme we will explore a few pages from now, high in the beautiful mountain country of Colorado.

CHAPTER SEVENTEEN

Home and Away

Boulder Colorado

I could really get addicted to this mountain air. Just like I did in the Italian Tirol, I find myself asking why I have never lived in the mountains, since the air at this elevation seems to suit me so well. This particular air fills the refreshing mountain country of Boulder, where I am spending a few days filming an episode of *Open Minds* with Regina Meredith of *GAIA*. From the moment I arrived in Boulder, just as I did in Val Gardena, Ortisei, I have had the strangest feeling of being safe and at home.

The feeling of familiarity is so vivid and yet I have never lived anywhere quite like it, with its quiet pedestrianized city centre, the friendly vibe of human-scale buildings, the plethora of eating places, the vibrant street music and the tempting array of artisanal stores. I love that even from the very centre of town I can always look to the beautiful backdrop of the foothills of Colorado's rocky mountains. Aspects of Boulder remind me of life in the city of Bath, England many years ago as a student of languages, and of our time as a family in the Yarra Ranges of Victoria, Australia. Yet I have never lived in mountain country quite like that of the rocky mountains of Colorado and I still can't put my finger on why it feels so much like home.

As I explore the streets of Boulder, I pause at the photographic displays memorializing the people, places and activities of the past which have helped define the character of the city and region. The freestanding rectangles of information are there to tell any interested passer-by, *"This is what happened here. This is who we were. This is where we came from. This is why the spot right here is a special place.* " Evidently, it is a profoundly human impulse to memorialize our shared experiences, from the ancient cave paintings of Lascaux to the selfie phenomenon today. When I visit more ancient cultures, either physically through the remnants of their built heritage or in the pages of their ancient texts, I frequently come across physical installations, the stone equivalents of these freestanding photographic displays in Boulder. For instance, the Bible records the ancient Hebrews' custom of erecting standing stones to mark significant places in their journey, identifying for the interested passer-by all the places of contact with the advanced non-human visitors who defined their early story. For example:

- Mount Sinai was adorned with twelve standing stones to commemorate Moses' world-changing encounter with the being Yahweh.
- In Bethel the patriarch Jacob erected standing stones to mark the place where he saw a hole open in the sky through which advanced beings (*elohim*) shuttled between the Earth's surface and somewhere beyond. (This is the story known as *Jacob's Ladder*.)
- Tel El Farah in ancient Samaria housed not only standing stones to denote a place of contact, but also a carved *naos* identifying the visitor who made contact in that place, and pointing to where it was that she had arrived from. The symbology of palm trees, bread cakes, a crescent moon and a cluster of stars tell us that the visitor was Asherah, and her place of embarkation, the constellation Pleiades.
- The Ziggurats of ancient Mesopotamia, the stepped pyramids of Central and South America and Cambodia
- The Temples of the Indian subcontinent, adorned with images evoking jet engines and flying Vimanas.

All totemize our ancestors' encounters with advanced beings and their technology. For our ancestors, exposure to visiting civilizations, even those who may have colonized and exploited us, represented an opportunity to observe technology more advanced than our own and wonder what it would be to have such capabilities for ourselves. Hence, we have depicted it in carvings, rock paintings and ancient texts and have done our level best to reverse engineer it.

In my previous book *The Eden Conspiracy* I spend some time describing a ceremonial device worn by Queen Xook of the Jaguar Dynasty in eighth century Yucatan. It was part of the regalia she wore to indicate her advancement and her ability to communicate remotely with the powerful feathered serpents of antiquity. Curiously it is an emblem which gives the appearance of a modern-day communications device, specifically a Bluetooth mic and earpiece. The fact that this was worn for the purpose of remote communication and in a location designated as *"the place of the voices"* ought to raise an eyebrow or two. These correlations cannot be coincidence.

The fact that Queen Xook's ceremonial accessory was in fact only a facsimile of something functional is given away by the accompanying bloodletting, an ordeal of a ceremony designed to create an altered state of consciousness, understood to be the necessary condition for achieving remote contact. The beings whom they expected to be in conversation at the other end of the line were the non-human beings who had ruled over the ancestors of the Jaguar people in the deep past, beings who their ancestors had seen to be adept with this kind of technology.

In this way the accoutrements of Jaguar priestcraft went beyond commemoration of advanced technology in the form of a cargo cult. It was a deliberate attempt to replicate the effect of what they had seen. So in a sense we can say that what the Pentagon calls

The Program in the twenty-first century is really nothing new. We have been trying to replicate our advanced visitors' technology for thousands of years.

As we saw in the previous chapter, paleocontact offered our ancestors more than crash courses in engineering. For instance, the Book of Enoch talks about the visitors helping us to have fun with more sophisticated clothing, with adornments and make-up. This is a much broader and more positive kind of cultural exchange than is suggested by today's language surrounding The Program.

Then there's that box in Kef Kalesi. With his bucket of technology in one hand and a pinecone in the other, Haldi's assistance involved a good deal more than a package of better medicines and foods, tools and toys. The pinecone hints at our brain's pineal gland. It tells us that Haldi's upgrade takes us beyond being better at mathematics and technical drawing. Pineal activation is about a fundamental and holistic shift in our levels of consciousness and intelligence. It is about our ability to engage with the powers and principles of the cosmos and tap a higher and deeper realm of knowing. If this is what interstellar contact meant for us in the past, then what might the future of contact hold for us?

Of course military intelligence will have a bias to assessing risk, and the Military Industrial Complex will naturally have a particular interest in what better armaments might be available today. It is their job, after all, to think in terms of strategic advantage. But what about the broader spectrum of advances that contact with an advanced civilization might have to offer? In an age of pandemics, auto-immune dysfunctions and adverse reactions, a little further support in medical science wouldn't go amiss. With the crisis of Greenland's melting ice sheets only accelerating, another helping hand in ecological recovery would certainly be timely. We are currently in an era when American society is witnessing a degree of violent polarization at an intensity not seen since the civil rights era of the 1960's and the Civil War of the 1860's. Internationally, we are witnessing heightened levels of xenophobia, wars and war-crimes, the bombing and occupation of others' homelands, and the denial of land rights, medical support or even water to indigenous peoples. As a result, the world is on the back foot in addressing the displacement of millions of families, who then face ill-treatment and internment camps once they have escaped the theatres of war and persecution. These geopolitical upheavals are happening at a level not seen since the Second World War. Perhaps another accelerated course in international civics would help the international community to create a better future for ourselves. If our cosmic visitors were able to support our ancestors on all these fronts in the past why would there be any less on the table today?

If the current generation of visitors is not hostile, and in eighty years it has been our forces shooting at them, not vice versa, then wouldn't it be helpful for our members of Congress to be considering the potential of broader and more positive lines of engagement? If we can be advantaged in our wider sciences, in our understanding of ourselves and our cosmos, in our ability to live peaceably and sustainably on our beautiful planet, isn't that a conversation worth pursuing?

Back in Boulder I am with my friend Matthew LaCroix. We have so much to discuss as we catch up on our respective research paths and explore the groundwork for joint projects in the year ahead. Today we are comparing notes on timelines for Lake Van, ancient Sumer, Urartu and the ancient Ararat Civilization, Haldi's box, and other intriguing artefacts in ancient Armenia. Matt has a case for pushing the timelines I have suggested even further back in human history, and we are challenging each other as we probe these questions together. But today is not all about work. Matt has generously devoted the bulk of his day to showing me the mountainscapes and stunning scenery of Boulder's surrounding country. It is so refreshing to be out in nature, surveying Long Lake, and we are drinking in the moment, photographing the breathtaking scenery and getting some real mountain air into our lungs.

As we climb higher into the foothills, it dawns on me why this place feels so deeply familiar to me. I may never have lived in the mountains, but seven hundred years ago, a place just like this was home to my ancestors. The family of my paternal grandmother, Florence Wallis, came from a mountain community high in the Swiss Alps. How had I forgotten that? The canton of Valais in Switzerland was home to the Valais / Wallis family for many generations, indeed for so long that the family and the place have the same name. So the lie of the land I am traversing in Boulder, in which the city sits in the valley where the major glacier once came through, embracing the city with the beautiful backdrop of the rock and snow of the mountains, laced with pine forest, meltwater lakes and streams; this was what home looked and felt like for my family for centuries on end. This environment is in my blood.

The change came in the 1300's when Europe was struck by a devastating epidemic. As it spread it became known as The Black Death, inexorably wiping out between a third and one half of Europe's population. It was in the wake of that devastation that my part of the Wallis family left Valais and walked the thousand-mile journey, across the Alps, across Germany, Belgium and France. Once in England they settled and made a new life for themselves as farmers. It was a life they enjoyed for the next four hundred years.

Of course, the Black Death was not the last ordeal the Wallis clan would have to contend with. From the sixteenth century until the nineteenth century, Britain's arable lands were progressively annexed by the powers of the day. Like the Mo'o and Anunu of Kam's ancestral story, the British crown enclosed what had been common land and privatized it, dispossessing my ancestors of their livelihood and forcing them to live more densely as tenants in the newly developing cities. So it was that my ancestors eventually found themselves living in the tight terraced streets of South London. Here, denser housing brought new challenges, health challenges in particular, and the seventeenth century saw another epidemic, the Great Plague of London which proceeded to wipe out more than a fifth of the city's population. Its progress was finally arrested by the Great Fire of London which obliterated a huge section of the capital city, incinerating more than one sixth of the remaining population's homes. The major civil engineering exercises which reconstructed London pushed its poorer residents into ever denser living conditions.

As if more stress were needed, the 1600s also brought the country two revolutions and a civil war. The century which followed ushered in a profound assault on public health through the foment of the industrial revolution which generated even denser forms of housing and years of life under a strict curfew. This was a time when econometrics was king and the watchword of nineteenth century business was to maximize profit through *"the depression of wages."* In this way the economic growth of the age was built on the calculated impoverishment of the labour force. The dependency of workers on their intentionally depressed wages was further reinforced through a long period in which the currency with which they were paid was progressively devalued. In a single year the British pound was devalued by 27%. From 1800 to 2023 the British pound lost 99.032% of its value. It is same draconic pattern we heard about in Kam's telling of the Mo'olele. Step one: move people off their land. Step two: force them to work for money. Step three: devalue the money with which they are paid. Hey presto, a managed population. Same as it ever was. To cap it all, the twentieth century then provided my ancestors with two world wars, four decades of economic depression and two decades of wartime privations.

In short, my ancestors on the Wallis side had to make it through a long gauntlet of challenges. When I review my Pa'an against the Hawaiian Mo'olele, every challenge which the Mo'o inflicted on Kam's Hawaiian ancestors was inflicted on my own ancestors in their seven centuries of life in Britain. Talk about history repeating! In that sense the stories of the Mo'o are really very relatable. Your ancestors may have lived through even greater challenges, displacements, slaughters and enslavements. The reason I enumerate all the trials and terrors of one quarter of my own ancestry is simply to make this point. You and I are descended from people who managed to survive every single one of those assaults. You and I are the descendants of survivors. Our Pa'an speaks of the survival wars and plagues. It is in our blood to adapt and thrive through even the most impossible of circumstances. This is a very good thing to remember next time you're feeling a little fragile!

I have other ancestral story I could share of my Ghanaian, Welsh and Viking ancestors. And you have your Pa'an too. If we were each to share the longer stories of our family heritage, I think it would encourage us to face the uncertainties of today with greater confidence and not fall prey to the continual broadcasting of fear and vulnerability.

If in the twenty-first century we really are in conversation with a diplomatic corps of more advanced neighbours, if Haim Eshed is correct and patterns of collaboration are already underway, then truly the stakes of our exopolitical diplomacy are high, right now and in the years ahead. Now is the time to be bringing our highest emotional intelligence to the table of today's *Council of Powers* if we want to see the best possible outcomes for humanity. Imagine the potential of our best creative minds allied with our best cosmic helpers. What new progress might emerge from a collaboration of that order?

There is really nothing quite like standing on a mountainside for enjoying a higher vantage. Here, all the anxieties and neuroses of the day to day recede into their proper perspective. I think my friend Kam is right. Being in nature like this really has a unique power to pull our minds out of their general funk and inspire us to bolder and more elevated ways of life. I make an inward commitment to conform with Kam's fatherly advice to me and spend at least an hour in nature every day when I get back home.

As Matthew and I begin our return journey from Long Lake to the city of Boulder, the gradual descent through the beautiful foothills of the Colorado Rocky Mountains is truly vivifying. It has been a great day as we have compared notes with each other, and our conversation has ranged over a lifetime of experiences. Somehow, I feel like I have come full circle, as if in some strange way I am carrying the memory of the Wallis family's seven-century journey from Valais to the present moment within my very DNA. I have to wonder how through the ages our ancestors were able to make it through such huge tectonic shifts. If I take our ancestral stories seriously then I have to consider the possibility that, at many times and in diverse ways, we had help – your ancestors and mine. In every generation, the lens of ancestral memory has allowed us to discern plural layers of influence upon the human story, some human, some non-human, some nefarious and some benevolent, and the message from antiquity is that every generation of hindrances has been met with a corresponding array of help. Guided by that assistance, my ancestors found a way to live and thrive, and so did yours, because ultimately our Pa'an is full of resourcefulness, nurture and ascension.

In the coming chapter we will certainly need the benefit of our ancestral wisdom. Nothing else will have prepared us for the unexpected revelations of the months and years that lie ahead. The question before us in the pages that follow is about how you and I will choose to respond to something new and unexpected as we arrive at the next rocky ridge.

CHAPTER EIGHTEEN

& CONCLUSION

Leaning Forward

Location Classified

We are high on an elevated rocky ridge, lying on the dusty ground, taking in an incredible sight. Below us is an expansive plateau, bordered by steep cliffs. We are overlooking a secure, enclosed zone, full of advanced technology. There are structures across the site constructed from materials we have never seen before and executed with a finesse way beyond the scope of terrestrial engineering. Within the site we can see a botanical zone, nurturing forms of plant life which will replenish poisoned and depleted soils, rebalance our ecosystem and bring health and equilibrium to our bodies. In one structure we can see a vast seedbank, a reservoir of DNA which will pull our endangered rhinos, tigers, elephants, gorillas, orangutans, blue whales and honeybees back from the brink of extinction.

On the site perimeter we can see advanced technologies which allow the mysterious visitors to fast-track colleagues and resources from a huge craft silently occupying airspace just beyond the clouds, almost but not quite cloaked. The non-human personnel staffing this open-air lab have a serene calmness about them as they interact and quietly go about their work.

If you haven't already guessed, we are surveilling a live action version of the ancient scene we saw carved into Haldi's basalt box. From this notional position you and I can spy on the activities depicted on the box, but as if in real time and from a clear vantage point. Now we see people moving among the mysterious visitors. These people are clearly human, yet even from a distance we can see they have the same peaceful and capable demeanour as the visitors. And they're beautiful. Something mystical has happened to them. They look like gods, an absolute picture of intelligence and strength.

We could stay here forever, quietly observing, except they know we're watching them. One of them has turned and is looking directly at us. Within moments a small tic-tac shaped drone is hovering silently overhead. We have been well and truly observed. Now they are watching us watching them.

So what do we do? Do we shoot the drone down? And if we can't, what then? Bomb the whole plateau? Furthermore, what do we do with the other enclosed zone we surveilled a thousand miles to the south? Would it be better to apply some *"preemptive defense"* and bomb that one too?

If this sounds like a tough call, let me volunteer another question: Did the artist who etched this scene into Anatolia's ancient basalt apply their incredible skill and artistry to warn us of the danger of alien invasions and takeovers, and prime us to launch a missile or drop a bomb should we ever witness a scene like the one he has depicted? I would suggest not.

The exquisite order and harmony of the design make crystal clear that the artist's intent was not to obliterate the wonder of what was seen in Haldi's ancient Armenia, but rather to preserve it. For the artist, Haldi's intervention was something to be remembered with awe, and the fact that it was so carefully engraved into the impossible hardness of basalt suggests that this message was intended to be viewed in ages long after the rise and fall of the Ararat civilization, by people who would sit and ponder it and perhaps understand its secrets.

If you spend any time in the world of ancient texts and indigenous narratives, you will quickly conclude, that our ancestors bequeathed their stories to us to help us understand the world around us. This is why I think it is vital that we consider the questions around today's ET presence in the light of our ancestral narratives – be they oral traditions, ancient texts or basalt reliefs. They give us a wider perspective and a longer view. If we choose to disregard them, we only rob ourselves of possibilities. We cannot allow the energy of mutual distrust and indignation between members of the U.S. Congress and the Pentagon to define how we frame the miracle of contact with other civilizations. We will only injure ourselves if we decide to engage with interstellar visitors on the assumption that any advanced visitor we can't shoot down must therefore be hostile.

By now you will have a shrewd idea as to what I think about the questions posed on the cover of this book. I suspect that, yes, we have already been invaded, and that through humanity's long story we have probably experienced many waves of colonization and annexation, some overt and some more covert. I am leaning to the view that the current iteration of ET oversight probably correlates with our period of rapid technological advancement, which began two centuries ago at the time of land enclosures, urbanization and industrialization, and then accelerated exponentially in the 1940's when UFO visits became impossible to disregard in the aftermath of our nuclear detonations. The dynamics of current collaborations may have brought both light and shade into our story, however, from where I am standing, the shortfall of humanity in our current geopolitics, suggests that a fresh review of any exopolitical alliances made eight decades ago is probably overdue.

Collating the official statements surrounding The Program which have been made in the last few years by representatives of The Pentagon, enough has now been officially acknowledged for anyone following the trail of news stories to know without any doubt that we are in contact. ET contact is no longer a matter of *yes* or *no*. It is more complex. It is about who we ally with on the one hand, and who, on the other hand, we may need to manage - with our allies' help. ET contact is not a matter of *for* or *against*. It is about applying scrutiny to the exopolitical decisions being made by those who currently represent us. I have an old-fashioned belief in democracy, an idea that things go better when the decisions affecting us all enjoy the light of scrutiny and the pressure of public accountability. A democratic society doesn't just mean a society that holds elections, it means a society where government represents the people rather than rules over them. It is a free and open society in which ideas can be freely discussed, and alternative courses of action can be considered openly and without fear. We need to extend this openness to everything that is happening right now in the world of exopolitics. As long as we, the people, refuse to be manipulated and mastered by fear of the unknown, this openness can only be to the good.

I want to be part of bringing what is hidden to light because I believe that knowledge is power. That is the very reason I write books. It is why I create documentaries and interviews on *5thkind.tv*. It is why courageous people like Haim Eshed, Ryan Graves, David Fravor, Chris Mellon, Thomas Monheim, Luis Elizondo, Alain Juillet, Tim Burchett, Andy Ogles, Nancy Mace, Marco Rubio, Chuck Schumer, Jacques Vallee, Garry Nolan, Eric W. Davis, James Lacatski and David Grusch are now publicly responding to ET contact in the way they are. Like me, they believe that accountability and openness will shift us out of the exopolitical dark ages into a bold new era. More than anything, our energy around this topic, whether we are researchers, campaigners or gatekeepers, needs to shift from anxiety and fear to one of intention and exploration.

Casting my mind back to the Congressional Hearing of July 26th 2023, I return to a pointed line of questioning issued by Congressmen Tim Burchett and Andy Ogles. Because I don't want to make this personal in any way, let me say that I completely understand where these gentlemen are coming from and I respect their sincere concern, but in the light of ancestral memory it seems to me that they could have asked very different questions and evinced exactly the same answers:

Question: "Are you *aware of any power in this world that has the kind of capabilities our visitors may have to offer us?"*

Answer: *"No. What our visitors have is far beyond any materials science that we possess."*

Question: *"Is it possible that these UAPs may be assessing humanity's current capabilities against our current needs?*

Answer: *"Yes." "Yes," "Yes."*

Question: *"Do you feel, based off of your experience and the information that you have been privy to, that our visitors could significantly enhance our experience as a species on planet Earth and as galactic citizens?"*

Answer: *"Potentially." "Potentially." "Definitely, potentially."*

Question: *"In the event that our encounters should prove to be positive and helpful, is there any way, looking to the future, that we could get on just as well without our visitors' assistance?"*

Answer: *"Absolutely not." "No."*

I am being playful, but at the same time I am making a serious point: Now is a time for us to be asking far better questions, deeper questions and a wider range of questions, if we truly desire a fuller understanding of what possibilities may be on the table for humanity, moving forward.

"So Paul, tell me, how do you see all this playing out?"

My pastor-friend, Will, has just binge-watched all the David Grusch material he can find and is now bending my ear via *Zoom* at my room in the beautiful St. Julien Hotel in downtown Boulder.

"How do you read it, Paul? Are we going to get some new disclosure, or is it all going to fizzle out with some Pentagon official waving a piece of paper in the air, telling us that some internal audit has looked into the blocking of David Grusch and has ruled that it was all totally legal and doesn't pose any real problem at all? Is that going to be how this story ends? Because that wouldn't surprise me at all."

I have to concede that Will's skepticism as to where this is all headed may be well-placed.

"All I can say, Will, is that I will be watching Washington closely for developments. The Program's new chief, Deputy Secretary of Defense, Kathleen Hicks, says she believes that 'transparency is...critical' and that she is 'committed to sharing AARO's discoveries with Congress and the public.'"

"To be honest, I can only laugh at a statement like that. The words 'transparency' and 'sharing with the public' have nothing to do with the work of the secret services. Let's be serious! What intelligence officer would believe for one moment that their job is to make their data transparent so that it can be shared with Congress and the general public? I mean really?"

On the other hand, it is possible that Thomas Monheim, the Inspector General of the Intelligence Community, allowed the David Grusch complaint to escalate in the way it did because he favours greater disclosure. At the very least it would appear that he is happy to draw public attention to what has already been disclosed.

DOPSR (*Department of Defense Pre-publication Security Review*) has also played a part by authorizing the escalation of public awareness. So much so that David Grusch is not the only person speaking about The Pentagon's possession of intact UFO craft with DOPSR clearance. In recent months Dr James Lacatski has spoken publicly about his work as leader of the Defense Intelligence Agency's iteration of The Program. He has made no secret of the fact that intact craft are in the possession of the DIA for *"exploitation,"* by which he means reverse-engineering. So the fact that, suddenly, we have two authoritative agencies within the intelligence community, DOPSR and ICIG both escalating the public conversation, that really is quite significant.

"There's a real push and pull around this in the U.S.A. Will, because while you have Chris Mellon, Thomas Monheim, David Grusch, Elizondo, Vallee, Nolan and Davis all leaning forwards, the opposite forces trying to push the genie back into the bottle are considerable."

"So, for instance, you remember the sixty-four-page amendment that Senators Marco Rubio and Chuck Schumer put together in 2023, the 2024 National Defense Authorization Bill? Well, by the end of the year, before it could become law, the two key provisions of their amendment were cut out of the bill."

The first key section to be excised from the amendment was a provision authorizing disclosure of ET contact events through a presidential panel. The second deletion was a provision to transfer possession of ET materials from the Pentagon's corporate contractors to the executive arm of the U.S. government. With both those sections removed, the Rubio-Schumer amendment was effectively neutralized.

"The key interventions which gutted the amendment which had, until then, enjoyed bipartisan support, were essentially to do with 'security concerns' which of course trump everything. The moment the name of 'National Security' is invoked by a Pentagon official, any entitlement to information whether by the U.S. government, or the public which elects it, is effectively negated."

"So, Will, having watched the rise and fall of that amendment, I am not so confident that legislation is really going to be the best route to disclosure. Where I am more hopeful is that with ICIG and DOPSR in a forward-leaning posture, we are far more likely to see authorities within the intelligence community increase the flow of disclosure we have seen in the last few years. I think we're going to see that really accelerate in the months ahead."

"Meanwhile, I hope that journalists like Ross Coulthart, elite figures like Chris Mellon, and politicians like Ted Burchett and Andy Ogles, Nancy Mace, Chuck Schumer, and Marco Rubio will continue to keep the pressure up. But if they are to do that, then you and I really need to support them, and signal to our media outlets that we want this kind of story pursued. We can't just sit back, scratching ourselves, if we want to keep this thing moving forward in a positive direction. We, the general public, need to make a noise around every disclosure that's coming down the track, and the way I read it, there's plenty to come."

"If this kind of action leads to greater transparency and new disclosures, then I can tell you a lot of people and institutions are going to be caught with their proverbial pants down – I mean governments, media figures, corporations, religious institutions, religious leaders. Their credibility at that point is going to take a hit, and for many I think we are already at that point."

My peroration prompts a quiet smirk from Will who, even from thousands of miles away, knows how to read my tone.

"I guess if more does get disclosed you'll suddenly look a little less like a flake, and a degree more credible! I get the feeling, Paul, that you would be happy to see the great and the good get caught with their pants down!"

I can't disagree with that. Taboo, official disinformation and gaslighting all deserve to be called out, whether it be from politicians, academics, or religious leaders. Who and what we are in contact with, what technology is available to us, and whatever else might be on the table for project Earth; these are matters for humanity, not just for a tight cabal of American politicians and corporate-types. These are things we should all want to know, and we all deserve to know.

"To be honest though, Paul, I'm not totally convinced that contact with cosmic neighbours would be a good thing. You seem to think it would be, but I am not so sure. I've seen 'War of the Worlds.' I've watched 'Mars Attacks.' Is contact with cosmic visitors really the wisest thing for the general public to be looking for?"

"If there really is something going on, don't you think it might be more complex than the average Joe is ready to deal with? People will go into meltdown, won't they? I mean, would you and I even know what to do with a close encounter if we had one?"

On this point I couldn't disagree more, though I think Will is probably goading me. He is the last person I would expect to blindly trust bureaucrats to make cosmic decisions for us in a spirit of pure altruism. However, I always like to take Will's provocations as an opportunity to press my point.

"Will, I take your point but, firstly, neither you nor I voted for the corporate-types who have trumped officials like Admiral Wilson and David Grusch in terms of access to The Program."

"Secondly, the most recent polls taken in the U.S. show that the around 67% of the population would respond to an official admission of ET contact by saying, 'Thank you! I had moreorless worked that out for myself!'"

"Where our culture is up to now, the mass panic rationale for secrecy is completely out of date. Just look at everything that has been revealed by the Pentagon since 2017. Contact with craft and bodies has been officially acknowledged and there is no panic."

"Thirdly, as for the merits of grassroots contact, you're a pastor Will. You know the Bible. Haven't you noticed that the New Testament actually carries a positive expectation of grassroots contact? The author of I John in the New Testament promotes encounter experiences without any of the fear you suggest is appropriate."

I think Will did not know that as he remains completely quiet at this point, which I take as permission to continue.

"If you look in 1 John 4, you'll see that the writer tells the early believers to expect contact experiences and intelligent communication with beings he calls 'spirits.'"

It's worth flagging that the author of *1 John* does not specify who or what these *"spirits"* are, other than to say that the reader can get helpful information directly from them. In ancient cultures the word *"spirit"* indicated a visitor with advanced capabilities in terms of travel. Such visitors could appear or disappear by mysterious means which the experiencer could not explain. For instance, the ancient Celts talked about the *Other Crowd,* who were able to arrive and depart *"in a beam of light."* On the other

272

side of the planet, the Yolngu people of Australia's Northern Territory spoke about the Mimi spirits as being able to arrive and disappear *"on the wind."* In the same vein, the New Testament book of Hebrews speaks about cosmic helpers transforming into *"winds"* and *"flames of fire."* So in no way does the word *spirits* nail down what kind of *non-human biologics* are being described to us.

What is clear is that the writer of *1 John* doesn't regard these *"spirits"* as transcendent or holy beings. Indeed he tells the reader not to be credulous, and to carefully filter whatever they might say, and to be ready to disregard any visiting entities who won't show respect to Jesus. Evidently, more important than pinning down who or what these non-human biologics are exactly, is the writer's concern that as potential contactees we should be sure to maintain our sovereignty and independence of thought. If as contactees we are willing to filter what we are told then apparently, we can expect to learn important new information from whoever or whatever these non-human visitors might prove to be. This is grassroots contact.

"1 John doesn't warn his reader against these close encounter phenomena, nor does he say that because the visitors might not necessarily tell you the truth you would really do better to hold up a crucifix and tell them to back off in the name of Jesus!"

The reason I say this is that many of my more religious correspondents tell me that is exactly what I should do if an ET visitor ever shows up.

"But big picture, Will, the writer of 1 John wouldn't be saying any of this if he didn't expect people to be having contact experiences with the potential of learning something that could be to our benefit as human beings."

273

Unfortunately, here I have to cut my conversation with Will short because I need to get myself booted and suited, and hopefully looking my best, to film at *GAIA TV* with one of my favourite interviewers, Regina Meredith. Within the hour I have been ferried to the studio and am sitting in hair and make-up, where the crew have found some potential for my facial improvement and are swiftly powdering me up for the camera.

While the necessary finishing touches are applied, Regina and I take a moment to compare notes on the themes we will be discussing on camera. Regina and I have met many times in online interviews, but this is our first time together in the same room. Regina is exactly the same off screen as on, and it is a real pleasure to meet her face to face. Moreover I am excited to be on her *Open Minds* show talking about my recent book exploring paleocontact in the Bible, *The Eden Conspiracy*.

Regina is a world-class researcher and broadcaster who has devoted decades to plumbing forbidden sciences and esoterica for their insights on human origins and human potential, and her research path has brought her to conclusions very similar to my own. The audience that Regina attracts at *GAIA* and on her own platform *reginameredith.com* indicates the huge level of public interest right now concerning questions of paleocontact, human origins and human potential. On my own *Paul Wallis* channel on *YouTube*, and at *5thkind.tv* we are finding the same public appetite with millions of hits on those of our documentaries and interviews which focus on these themes. As we continue chatting, I find that Regina and I are singing the same song.

"We have had control structures on Earth for a long time imposed by beings who do not particularly have our best interests at heart. And the same influences are interfering with humankind today, would that be fair to say, Paul?"

It would be fair to say. At least that is my conclusion after my world tour of ancestral story.

"So, on the one hand we have had all those control structures imposed on us. But wherever there is that there are also the countervailing species who come, species who wish to be of help because we are such a unique population. It's such a persistent story throughout history and around the world. There are many benevolent others who are here to help us remember that."

I am in total agreement, and for me this is the order of the day. Given what has already been disclosed by the likes of Dimitri Medvedev and Haim Eshed, the question should no longer be, *"Are we in contact?"* It should be, *"Who are we doing business with? Who are our allies and are we engaging their best support? Do the factions involved in our covert collaborations have the best interests of humanity at heart, or might we need to extend a hand towards others of our cosmic neighbours?"*

Like me, Regina is confident that we have benevolent company at hand every bit as much in the present as in the past and, like me, Regina finds reasons for confidence in the deep memory of contact carried in ancestral story.

"Humans are so unique. We are a species with a physical, dense body, able to demonstrate the attributes of love, harmony and sustainability. Many species are watching us for that alone."

Regina's enthusiasm is palpable.

"I think we are starting to change as a species and are beginning to wake up. The great virtues, care, compassion, truth, love and beauty, all archetypes in themselves, that's what we are reawakening to. With those we can throw off chains that have bound us for thousands of years."

"When we learn to develop wisdom and develop our care for one another and for this beautiful intelligent planet we live on, when we can learn to be wise and mature and live in harmony, and when this drops down into our hearts, and into the way our mind actually works, this will be the holy grail."

I couldn't put it any better. It's the same vision Kam unveiled in his telling of the Mo'olele back in Hawaii, and it's a message which threads through many of our world's indigenous stories of human origins. In those narratives, I recognize a shared intent: to reveal the outline of a possible future, to reassure us and persuade us that we have it within us to be more than we are. This is what drives me in my research today, the hope that when we truly understand who and what we are, we can begin to create a better human experience.

Having shared this journey with me, you will understand why I believe we are on the cusp of something huge, an existential shift for humanity. Variously observed, adapted, exploited and assisted by the many who have visited us, one way or another the human race has been an ongoing project of Ed Mitchell's *Community of Space-faring Civilizations* for a very long time. If he and Haim Eshed are on point, then our relationship with that community will change in the moment when we finally master spacetime and become a truly space-faring civilization. In that instant our relationship with the cosmos will change because we will suddenly have access to it. The powers, people and principles of the cosmos will be available to us. We will be in each other's midst, not as retrievals, abductees or resources for research, but as neighbours. If I am right about that then corporations like Lockheed-Martin, Skunkworks, Northrop Grumman, Boeing and Raytheon are at the very cutting edge of humanity's existential shift. In this epic endeavour the public success of our aerospace technicians is all our success, because mastering this next chapter in our materials science will change everything.

Yet we have one last obstacle to overcome, and it is a significant one. Corporations exist for one thing and one thing only, to generate as much profit as possible for their shareholders. If a corporation's researchers, working on the basis of open-ended, unaccountable, black budget funding, suddenly achieve their goal and break through to the desired new technologies, the financial incentive driving that corporation will be in the direction of releasing that breakthrough in as gradual and atomized a way as possible. I am not saying this to insult the integrity of any individual. I am simply pointing out that the Pentagon's decision to privatize The Program has created an incentive which runs counter to the hopes of many who would like to accelerate disclosure and move this whole thing forwards. It's a *Catch 22* situation.

Yet the story doesn't have to end this way. It just needs people like you and me, citizens, journalists, broadcasters and political leaders, to shine all our democratic scrutiny on The Program and its gatekeepers and make the issue of public accountability part of mainstream conversation. Ultimately, every one of us is a stakeholder in The Program's success or failure, you, me and every human being on the planet. Now is the time for us to push together, and push past this vital tipping point in the story of humanity.

It's not every day that I feel this bullish. The forces at play can seem overwhelming and intimidating and can leave onlookers like you and me feeling disempowered and totally irrelevant to how things play out. What is the point of me throwing my hat into the ring? I am not a politician or a lobbyist. I am not a ufologist. I am the paleocontact guy. Who am I to speak into such a fraught subject? In my moment of hesitation, this morning, it is my friend Laura Eisenhower who gives me the nudge I need.

"Paul, you have written an amazing series and with what timing on this particular topic! Everything in our past has led up to what we are experiencing right now and we need to be aware and make mindful choices. It is the most important thing right now."

Denver International Airport

"If they delay any further, I'm worried that I'm going to miss my connection at O Hare."

I am hurriedly texting my wife Ruth, crouched over my phone in the least crowded corner of a noisy departure lounge, doing my best not to let the tightening timeframe stress me out. I am missing my family and I like to keep them apprised from day to day of my progress around the world. It has become a touchpoint for Ruth and me to exchange texts whenever I am sitting in a departure lounge, waiting to embark on the next leg of another global tour.

"How's your foot Sweety?"

To be honest, I had almost forgotten about my spider-bite.

"Oh, it's great thank you! Kam's ash paste has worked wonders. He's really onto something. My foot's back to its normal flat fish shape! In fact I am fit as a fiddle and fully functional. I'll have to show you when I get home!"

The strident voice on the tannoy cuts across the hubbub of the departure lounge, interrupting our momentary reunion.

"Passengers for flight LH7465 to Chicago, O Hare International Airport, we will begin boarding in approximately fifteen minutes. Please have your boarding passes ready for inspection."

As a child, I delighted in following my dad's jet-setting, and many times we had the pleasure of traveling as a family as his work spanned the globe to keep our skies a safer place. Today my own children are making a similar journey with me as I also look to the skies and explore the planet, although with other species of traveler in mind. On this particular trip though it's just me, and after two weeks away from home already, I am eager to get back to Ruth, Ben, Evie, Caleb and the cats. I don't want my studies of ancestral wisdom to be theory only, and as a family we are doing our best to keep learning all we can and grow in our ability to nurture an *open manawa* and attain the best health and *na'au* that we can. Meanwhile, today's flight has me headed for another country. Chicago is just a connection.

In fact this journey is going to be a long one, three flights, with nineteen hours of flying and connecting, followed by fifteen hours on the road. In all it will be at least a full day and a half before I am finally ensconced in my next billet. *"Passengers in groups one and two, please now prepare to board the aircraft.*

Though I am traveling alone today, my arrival at the other end will put me in the company of four of my favourite fellow-researchers. It's going to be an adventure for the whole team, and I have done my best to build up my muscle strength and stamina for the physical challenge of it. I am very aware that the youngest on the team is a full twenty years younger than me, and I am hoping that my commitment to the gym during my weeks on tour will have readied me to match the pace of my juniors on what's going to be a grueling expedition.

Our destination is an exciting one, full of hard copy data from a long-lost civilization, data which has the potential to revise our whole understanding of human history. Together, we will be probing archaeological sites on opposite sides of the planet which have encapsulated a critical pivot in the story of humanity, one which occurred the last time our planet was overtly invaded. And, because you have done me the great honour of travelling with me to this point, it's an experience I would like to share with you.

"Passengers in Group 3, we now invite you to board."

O.K. That's us. We're on the move. If you choose to join me, we will soon be back in the lands which once were Upper Mesopotamia, examining certain strange artefacts which have lately come to light at Kef Kalesi and Ayanis Kalesi. I want to know what really happened there, and for that reason I want to dive deep into a layer of information which I may have missed in our previous foray into Anatolia. In particular I would like to analyze more closely an enigmatic collection of objects which, thousands of years ago, in an age before every religion and in a time long forgotten, laid the foundation of every science and every one of the world's wisdom traditions. I am going back to get a closer look at that box.

Made in the USA
Las Vegas, NV
27 April 2024